T0330752

Moving from Project Management to Project Leadership

A Practical Guide to Leading Groups

Industrial Innovation Series

Series Editor

Adedeji B. Badiru

Department of Systems and Engineering Management
Air Force Institute of Technology (AFIT) – Dayton, Ohio

Kansei/Affective Engineering
  Mitsuo Nagamachi

Kansei Engineering - 2 volume set
  Mitsuo Nagamachi

Learning Curves: Theory, Models, and Applications
  Mohamad Y. Jaber

Modern Construction: Productive and Lean Practices
  Lincoln Harding Forbes

Project Management: Systems, Principles, and Applications
  Adedeji B. Badiru

Research Project Management
  Adedeji B. Badiru

Statistical Techniques for Project Control
  Adedeji B. Badiru

Technology Transfer and Commercialization of Environmental Remediation Technology
  Mark N. Goltz

Moving from Project Management to Project Leadership

A Practical Guide to Leading Groups

R. Camper Bull

CRC Press
Taylor & Francis Group
Boca Raton London New York

CRC Press is an imprint of the
Taylor & Francis Group, an **informa** business

CRC Press
Taylor & Francis Group
6000 Broken Sound Parkway NW, Suite 300
Boca Raton, FL 33487-2742

International Standard Book Number: 978-1-4398-2667-6 (Hardback)

Library of Congress Cataloging-in-Publication Data

Bull, R. Camper.
    Moving from project management to project leadership : a practical guide to leading groups / R. Camper Bull.
        p. cm. -- (Industrial innovation series)
    Includes bibliographical references and index.
    ISBN 978-1-4398-2667-6 (hardcover : alk. paper)
    1. Project management. 2. Leadership. 3. Executive ability. I. Title. II. Series.

HD69.P75.B85 2010
658.4'092--dc22
                                                  2010010206

**Visit the Taylor & Francis Web site at
http://www.taylorandfrancis.com**

**and the CRC Press Web site at
http://www.crcpress.com**

Dedication

To Doctors Robert J. and Vivian A. Bull,

Who daily show the world what true leadership is

Contents

Section 3   Getting Things Done as a Project Leader

Foreword

Many books have been written about project management, but this is the first I have seen that focuses on project leadership. Likewise, there are many project managers but very few project leaders. The few leaders who truly lead with vision, inspiration, empowerment, and respect make a huge difference in the work they do. As you read this book, it becomes clear that the thoughts and real examples shared in this book will be useful to current and future leaders.

In my career I have worked on thousands of projects. Many of these projects went well, and some not as well as I had hoped. Through these experiences, I've learned that projects' fortunes do not hinge on issues with other people. Instead, the key to success is often the adept management of information.

In an information-rich world, and especially in the information technology industry, information is so ubiquitous it can almost be overwhelming. I believe the key to successful projects, and therefore successful organizations, comes from the ability to skillfully use information, along with understanding, motivating, and leading people. Along these lines, I believe one aspect of a great leader is also the ability to practice servant leadership. I became aware of servant leadership almost twenty years ago and have personally seen the difference it makes when leading a team. This is especially true when the team is asked to do extraordinary things, because its members will follow a servant leader much more effectively than a dictatorial leader. A servant leader listens and enables the team to do far greater things than the members could do themselves.

A great leader also knows how to care for all constituents: customers, employees, owners, and other stakeholders. This requires a unique and broad perspective rarely found, but powerfully demonstrated by a handful of strong leaders who get results. The newest members of the workforce, those within the Generation Y and Millennium generations, bring new ways of thinking and challenge past leadership models. This book explains how a leader works within these new dynamics and also shows how leaders throughout history have embraced leadership principles to make a difference.

In today's demanding, fast-changing world, success comes only to those who can lead effectively and get the most from their team. I grew up in a small town in southern Virginia and started a small lawn care business. These early lessons in running a small business proved valuable later in my career. No matter how large or small the task ahead, having an entrepreneurial mind-set from the start and finding a way to get things done with the resources and network you have will yield great results. This book helps you find that entrepreneurial perspective. Setting high expectations and holding yourself and your team accountable will help you, as a project leader, to be successful.

Finally, the book includes exercises that let you truly experience what it takes to be a great leader. Once you've finished this book, you'll know what it takes to be a better leader at work, at home, and in other aspects of your life. So buckle up and enjoy the ride. I think you'll walk away with a new perspective on project leadership.

John Hinshaw
Kitty Hawk, North Carolina, 2009

Acknowledgments

As with any sort of book, no one person can be credited with all of the work. The following is just a brief list of the individuals who have either wittingly or unwittingly helped me with this document. These are the people who have influenced me, pointed me in the right direction, or followed me, believing that I knew what I was doing.

Cheryl Badger: For deep insights and always making good suggestions.

Judy Balaban, PMP: For first presenting the idea of this book, always being a great sounding board, and allowing me to be the straight man.

Robert Balascio, MBA: For showing me that leadership in the corporate world can be aboveboard and inspiring.

Tierney Bryant: For the wonderful stories and deep insight.

Alexandra S. Bull, MD, MBA: For all that you are and all that you are to me.

Robert Coultas: For encouraging the positive and always being One Voice.

Scott R. Crowther: For agreeing to try some unconventional thinking when none of us knew if it was going to work. From a quiet conversation on the bench till now—what an amazing journey.

Christopher deVinck, PhD: For teaching me that the personal story is the most powerful, and using your life story to change the world.

G. Zachary Edelstein, PhD: For all the feedback and sage advice.

John Hinshaw: For believing in this book, and for the good times and insightful conversations around the world. I'm looking forward to more.

Yana Kropotova: For being you and making me strive to be better every day!

Kimberly E. Phillips, MEd, BCBA: For keeping me on track and on schedule, and dealing with my eccentricities. You are greatly appreciated.

Robbie Robertson: For being a true Southern gentleman who displays servant leadership every day of his life.

Olin J. Storvick, PhD: For being one of the twelve men and explaining to me the idea of the shadow of the leader.

Barbara A. Verde, PMP: For being a good friend and confidante—and, of course, the creator of the Big Stick/Warm Blankie.

Jeffrey Wells, AIA: For trusting me enough to allow change within one of his favorite institutions and for saying quietly to me, "The greatest teachers teach by telling stories."

Kimi Ziemski, PMP: For your indomitable personality and great conversations that pointed me in the right direction.

About the Author

Camper Bull, project management professional (PMP), a cofounder of Armiger International, has held leadership and management positions in both commercial and volunteer environments. He has led a division of an international software company and served as lead account executive for Fortune 500 companies at both Genigraphics and ACI. Camper led development of the ACI multimedia marketing product team and developed its corporate image as an industry leader and innovator in the market. He managed the development of new products, including the launch of wireless paging at Bell Atlantic.

His background in developing the strategic planning and growth initiatives for VISTA Computer Services, including restructuring the company to better address the market with a global offering, has significantly helped the company gain a global footing in the software industry.

He has developed and delivered several leadership training programs in business and other industries as well as for high school seniors and foreign exchange students. Camper is the author of a significant PMP Prep program for the purpose of helping project management professionals prepare for the PMI certification examination, and is on the development team for PMBOK version 4. Proficient with computerized business applications, Camper has been able to transfer these practical skills into useful training tools, helping attendees gain a fast yet comprehensive understanding of these tools in their business environments.

His extensive travels throughout Europe, Africa, and Asia have provided an additional strength to the many abilities of communication and project management training that Camper brings to the platform.

Credentials

As an active member of the Project Management Institute and a certified PMP, Camper has also been recognized by Rotary International with the Paul Harris Fellow Award for furthering human relations among peoples of the world, and by Bell Atlantic Paging for leading the project teams that developed and implemented the largest amount of successful product in Bell Atlantic's history.

Camper was educated at Drew University, Madison, New Jersey, in economics with concentration in complex organizations, business ethics, managerial economics, and finance. He completed additional studies in the European Community in Brussels along with intensive French language studies in Paris.

Introduction

When I was a young boy, my family lived in the Middle East near Jerusalem. My father was an archaeologist, and at the time he was excavating a site called Caesarea Maritima, the seat of the Roman Empire in Palestine on the coast of the Mediterranean.

One Saturday afternoon, we were preparing to have tea in the garden of an archaeological institution known as the Albright School of Oriental Research when three gentlemen arrived to speak with my father. For lack of anything better to do that day, I went with my father to greet his guests.

When we reached the lobby of the institution, two enormous forest green doors swung open. In the doorway stood the three gentlemen, one in a gray suit with a white shirt and a spectacular purple tie. The other two were slightly taller, wearing black jackets, black vests, and black hats.

I knew from the men's garb that they were members of an ultra-orthodox group. Two men were Hasidim and spoke only Yiddish; the other man was the representative of the head rabbi of Jerusalem and translated for the Hasidim. My father is a good Southerner from Virginia, so he invited the men in to have tea with us.

We quickly discovered that the visit was not a social call; the men had come to accuse my father of desecrating Jewish graves.

My father and his team were excavating within the city walls of Caesarea Maritima, and the burial sites the men were referring to had been marked with Christian lamps—an improbable memorial for a Jewish grave and more likely the final resting place of Crusaders from the First or Second Crusade. My father showed the men pictures of the burial site and said, "Tell me about Jewish law. Is it not true that you do not bury your dead inside the city walls? We are excavating inside the city."

The pictures and his explanations fell on deaf ears. The Hasidim said, "You are excavating Jewish graves and you will stop immediately."

My father explained that he was licensed to excavate by the State of Israel, to which the Hasidim replied, "We do not recognize the State of Israel."

The exchange continued for several moments, until one of the men finally said, "If you do not stop excavating our people's graves, I will bring one hundred men to your site and stone you to death."

My father apologized again for the men's concerns, but firmly restated his position: he was licensed to excavate and would continue to do so. The men shook their heads and left, but the Hasidim issued a final warning before departing. They told my father, "Tomorrow we will come and stone you."

After hearing this threat, I insisted on going with my father to the site. He agreed to allow me to accompany him, so long as I followed two rules. Rule one, he said, was "Raise no hand." Rule two: the only thing I could have in my hand was a camera.

Obviously, I had several other ideas of things to hold in my hand if men were coming to stone my father, and a camera was not anywhere on the list. I'd grown up with media images from the Western world and was under the impression that my best option to save my father from these men's violent threats was to revert to violence as well. Unfortunately, I had been forbidden to do any such thing and, at that age, my father's word was still law.

Early the next morning, we traveled to the excavation site. Once there, my father stood up, explained the situation, and asked everyone to leave the site for safety. Twelve volunteers flat-out refused to leave, and in spite of his protests declared that they would stand with him no matter what.

At 8 a.m., a family friend from Israel informed my father that several tourist buses had left Jerusalem that morning and were heading our way. At 10:30 a.m., four large white buses pulled up at the site. Masses of men in black hats, white shirts, and black jackets streamed out of the buses, and, as they stepped out, they each reached down and picked up a large stone.

Our only protection from the mob of stone throwers was three strands of barbed wire. A sign bearing the words "Please keep out" in four languages hung from one of the fence posts. I stood on a hill near the site, a place where I normally sat to watch the waves of the Mediterranean. On that morning, though, I only had eyes for the sea of black hats as they moved closer and closer to the gate.

As my father moved toward the sea of black, I looked helplessly at my hands to examine my only defense weapon: a camera. I watched with trepidation as my father walked up to the first man he met. He extended his hand and waited. The other man was holding a rock in his hand, and do you know what he had to do to shake my father's outstretched hand?

He had to drop the rock.

I watched my father move through the sea of black and shake each and every man's hand, one by one. One by one, the men dropped their rocks to shake hands with my father. At that moment, I was awestruck by the power of an open hand, the power of a shake, and the power of getting to know an individual on humankind's basest level.

The actions and choices made that day were different from anything I'd ever witnessed before and forever changed my definition of leadership. My father made the decision to allow me to watch his possible death. The men who came to the site with hate in their minds left with peace in their hearts, and a new sense of understanding.

Perhaps the most incredible to me was the fact that twelve of the archaeologists who had worked with my father for many years chose to stand with him that day, even though they knew they risked death or serious injury as a result of that decision.

How did my father inspire his team to do something so above and beyond what was expected?

I believe those volunteers made the choice to stay behind because my father was their *leader*, and had always been a wise and giving leader—a servant leader. My father had always given himself to his men, so when presented with the opportunity to repay the courtesy, these volunteers offered up the greatest sacrifice they could muster—their lives.

Management versus Leadership

> Inventories can be managed, but people must be led.
>
> **—H. Ross Perot**

The premise of this book is that companies need more project leaders and fewer project managers. Project leaders lead projects in much the same way that successful entrepreneurs lead companies.

Whereas project managers can "manage" a project by making decisions about organization and tasks, project leaders go beyond these tasks by incorporating an overall vision and dedication to the team. The result is that project leaders are more innovative, more passionate, and better able to overcome difficulties and challenges.

Let me flesh out this difference a little more. To understand project management, it helps to understand where the word "management" comes from in the first place.

During the Industrial Revolution, there was a dramatic shift in the way that people worked and lived. People went from living in small groups (often extended families) and making their living mainly by subsistence farming to living in large cities and working for large organizations. These organizations, often businesses or conglomerates of businesses, needed to develop tools and techniques to get relatively mind-numbing work done as quickly and efficiently as possible.

Out of this need grew the idea of management. In the beginning, management was developed to control individuals who were simply putting in their time and exchanging dollars for hours. Management techniques were designed to maximize individual functionality, and each individual was viewed as replaceable (as were all other parts). Many of these techniques revolved around measuring productivity and increasing efficiency so that more widgets could be produced in a shorter period of time.

These methods, by their very nature, categorized employees as "things" rather than human beings. These management techniques are still pervasive throughout our system. For example, if an accounting firm buys a piece of equipment today, the purchase is considered an investment. A newly hired accountant's salary, on the other hand, is considered an expense.

There are times when some of these dated management methods are appropriate; however, there has been another transition in the business world—a paradigm shift, if you will—and many of these techniques are simply obsolete. Today, there is much more to leading projects than just making sure that your employees show up and complete their work on time.

Let me bring home just how obsolete the old mind-set is. Around the same time that these management techniques were being developed, leeching—that is, draining the blood from one's system by attaching leeches—was still a common and respected practice in medicine.

Now imagine if we were using the same medical techniques today that we had used during the Industrial Revolution, including the practice of bloodletting by using leeches. Doctors would not be talking about germs or genes, and certainly would not be prescribing as many medications or therapies, but you might hear leading hospitals making statements like the following:

- Our leeches are only from the finest organic farms!
- Our state-of-the-art bloodletting rooms are designed for your comfort as well as your health.
- Our bloodletting process has been rated number one by JD Powers & Associates.
- Our drive-through bloodletting shop has revolutionized care.

Luckily, modern medicine was able to move beyond the leech-and-bleed model. We now know about germs and genes, and I'm certain that our medical practices are better for our current knowledge. Once the scientific community discovered that microorganisms were creating infection, the entire medical community changed its perspective on the best way to treat people.

If the medical community has evolved in the face of new knowledge, why do organizations and corporations cling to management techniques that are just as obsolete? Most organizations continue to manage people in much the same way companies managed factory workers during the Industrial Revolution. While the medical field has shifted its perspective in light of new information, knowledge, and societal needs, too many businesses have failed to make the leap and begin recognizing individuals as an important part of the whole operation instead of as easily replaced widgets.

The world has moved on. The workplace is more diverse, more varied, and filled with an entirely new set of challenges. There has been a "sea change" in the way that we do business, and now is the time for organizations to catch up and evolve to a more effective style of leadership and project management. Treating individuals as worker bees and replaceable objects simply will not work in the modern business world.

The Case for Communications

If project leadership is the main thesis of this book, the power of effective top-down communication is the tune you can hum after putting the book down.

In the old Industrial Revolution model of management, a supervisor gave one a task (or set of tasks) to accomplish by some arbitrary time or date, and then waited for the results. Employees were hired for their experience and were expected to do the tasks asked of them and nothing more. Authorization for almost everything came from the top, where the most experienced employees were (supposedly) promoted to positions of responsibility.

We've all experienced the consequences of using this outdated model. Employees will seldom, if ever, take responsibility for mistakes. Finger pointing and "buck passing" are common. Any semblance of a problem or challenge stops progress until management can be consulted; employees who take independent action are often penalized for doing so. Territorial fights break out between departments. Forms and memos replace anything resembling actual human communication. I'm sure you're familiar with the drill.

The project leaders of today try to prevent this horror scene by overturning the old methods of management. They do this by transmitting what some have called "organizational clarity."

Chances are that you have run across attempts at organizational clarity already. Rows of books and CDs about business have touted the need for clear values, objectives, and goals for years. You have probably heard about mission statements or vision statements (and might have even worked on one yourself at one point). Small armies of consultants speak the lingo, often leaving inspirational posters and vague-sounding mission statements in their wake. Sometimes we even hear stories about these things working!

Nonetheless, organizational clarity is not about summing up your business in a few sentences, or speaking the right number of buzzwords. Choosing the right words is important, of course, but organizational clarity goes beyond finding a neat and tidy description by which others can get to know your company.

The kind of clarity I'm talking about is when an organization's leadership defines and agrees on the fundamental ideas behind the company and communicates those ideas effectively to everyone within the company.

Organizational clarity, when done correctly, gives employees at all levels of your organization a common vocabulary and a common aim. When employees have these tools, they make fewer assumptions about what needs to be done. Eventually, they become able to decide for themselves what is important and what is not, and will rely less and less on management to make this distinction for them.

When employees are able to make decisions for themselves and decide what actions are best for the company, interesting things happen. They take advantage of their ability to make decisions on their own, handling more details more quickly and freeing up the project leader's time to do what a project leader does best: lead. They also become more likely to resolve problems. Of course, they might make more mistakes in the process, but they will fix these issues faster, and are more likely to take responsibility for their mistakes and learn from them.

By paying close attention to organizational clarity and the way messages are transmitted within your organization, you will find new ways of empowering people while increasing efficiency—something the old management style can rarely boast.

The Power of Empowering People

Now that I've mentioned empowering people, I should say a little bit about what this means and why empowerment is a good thing.

"Empowerment" is one of those buzzwords that grew out of the business culture of the 1990s, and the theme of "empowering people" has appeared in almost every book on leadership since. But how exactly do you empower people? And, perhaps more importantly, why should you?

Let's talk for a moment about power. Power is simply the ability (or resources) to get people to do things because, in the backs of their minds, they understand that you mean, "Do this or else." Power uses threat, punishment, and bribery to motivate (and I use that term loosely here) people to do things—often things that the people would not want to do naturally on their own.

Power

- comes from many sources: authority, affluence, arms…and that's just the "A's."
- can manifest itself in many different forms as well: financial leverage, executive authority, favors owed and accumulated, and even just brute force.
- can be seen in many different places: parents have power over children. A boss has power over his or her employees. Teachers have power over their students.

Our society is impressed by power—some people are even obsessed with power. And it is no wonder: power can deliver some impressive results sometimes. However, just because an individual or an organization has power does not mean that the individual is effective, or that the organization is healthy.

There is good power, and there is bad power. The Mafia has plenty of power, if you think about it—but most of us wouldn't want to run our businesses like the Mafia.

Whether good or bad, power tends to be concentrated in the hands of a few people at the top: the king, the general, or the CEO. The problem with this is that everyone else in the organization spends his or her time and energy in attempts to get more of that power for themselves—or in trying to avoid the wrath of the person at the top wielding the power.

Unfortunately, while everyone is vying for power, no one is focusing on implementing the organization's original mission or taking care of the customers; they lack organizational clarity and wise leaders to follow.

True leaders don't try to get the upper hand in the power game—they refuse to play the game altogether. Rather than trade in power, true leaders trade in personal authority. Personal authority comes not from a title or job description, nor from a mandate or a paycheck. Personal authority comes from influence, from the skill of getting people to willingly do what needs to get done by working with them, not through them or above them.

The chasm between power and personal authority is huge once you begin to explore this contrast. Power, for example, can be bought, sold, given, or taken away, but personal authority cannot be transferred in this way. Personal authority must be grown and accumulated by personal relationships, and by using techniques that make people want to work with you and for you.

Personal authority also comes from different sources than power. Personal authority grows out of trust, competence, and a reputation as a problem solver. The leader who has the trust of his or her team, who is competent in running and demonstrating its practices, and who garners a reputation as a problem solver is the leader that everyone will want to work for—and work with.

My father didn't get his volunteers to stand by his side by exercising power. In fact, he demanded that his men leave the excavation site that morning. My father certainly did not force the men to make the decision to stay; rather, he influenced them to stay. The volunteers willingly did what needed to be done because of my father's influence and authority.

"Empowering people" is just a catchphrase for the main method of exercising personal authority. When we empower people, we give them the tools and resources to do their jobs and to solve problems—both with minimal interference.

When we give our frontline employees power, we put the most responsibility in the hands of the people with the most information and the best skills for getting a project done. Period. In an ironic twist, we end up increasing our own personal authority by giving our team members and employees all the power.

I invite you, then, to hold back any gut reactions you might have to phrases like "empowerment." The word is overused, but the idea behind the word is solid. To be a leader, you must learn to give power to your employees and get used to the idea that you serve them, not the other way around.

Engage in the Act of Leadership

As you can imagine, serving your followers requires you to engage and interact with them on a deeper, and at times, a more challenging level. If we examine a continuum depicting a leader's level of involvement in the project management process, we can see how an "active" leader appears to achieve a happy medium between a destructive and an engaged leader.

In today's business world, however, achieving balance between the two poles is not enough. Whereby "active" was acceptable and even encouraged in the past, project leaders now must move from an active and integrated role within the team to become a truly great project leader. An engaged project leader knows how to hold people accountable, how to be assertive, and how to be honest without alienating. An engaged project leader will be able to convince the entire team to stand behind their mission, without individual members agonizing over perceived sacrifice or striving for personal recognition. Under the guidance of an engaged project leader, teams will own the project, commit themselves fully to achieving the desired results, and take accountability for their actions. In other words, the team itself will become engaged.

Leadership

Destructive   Passive   Active   Integrated   Engaged

FIGURE FM.1
The leadership continuum.

The leadership continuum (see Figure FM.1) allows us to talk about leadership as a positive process. As a project leader, you must move individuals on the team who fall on the destructive side of the leadership continuum into the active side by engaging team members in the process, helping them own the project, and proactively working toward the agreed-upon goal. Engaging as a leader and getting your team to do the same requires not only strong leadership, but also an understanding of how leadership is developed.

The single biggest responsibility of a project leader is to move people who are destructive, passive, or even active to become more integrated and engaged. By definition, a leader's role is to lead his or her followers to a higher level. Every individual member of the team must be pushed, prodded, encouraged, or led to become more engaged in the project. Studies show the more engaged people are in their work, the more effective they are in terms of performance and productivity. If project managers can move teams to an engaged project process, they can create truly high-performing teams.

Consider again the leadership continuum pictured in Figure FM.1, but this time use the graphic to depict the level of involvement of each member of your team and the team as a whole. Where does each team member fall on the continuum? Where does the team as a whole rank? Now consider how each of the team members and the entire team could become more integrated and engaged in the project.

President John F. Kennedy engaged the entire nation when he issued his famous lunar challenge to the United States. Kennedy said, "We'll put a man on the moon and bring him back safely within ten years," and the entire country was captivated—and committed to achieving this lofty goal.

Several years after Kennedy laid down the gauntlet, a consultant was walking down the halls of the NASA headquarters when he noticed a gentleman mopping the floors. The consultant decided to test what he had heard about NASA's extraordinary commitment to Kennedy's goal, and he asked the man what he was doing. The janitor looked up, and without missing a beat replied, "I'm putting a man on the moon within ten years and bringing him back safely."

From the janitorial crew sweeping the hallways to the crew manning the mission control room to the astronauts who took the first steps on the surface of the moon, every person involved in the space program took complete ownership of their mission—and just look at the results.

Kennedy's words engaged the minds and eyes of the world. An entire industry immediately sprang up to further his objective. New thoughts and new ideas, such as Velcro, memory foam, cordless drills, freeze-dried food, and the cell phone, emerged as a result of the United States' quest to put a man on the moon. By engaging his team and uniting the country behind a common purpose, Kennedy was able to lead the United States to achieve things that had never been dreamed of before—and the world has never been the same since.

How to Use This Book

In the first half of this book, you'll learn the theories and practical knowledge required to be an extraordinary leader; what it is, exactly, that you need to do to be the best leader you can be. In the second half of the book, you'll find the tools and processes you'll need to put that knowledge into place.

This book is supported by a multimedia Website. Go to www.managementtoleadership.com.

Section 1

The Case for
Project Leadership

1

The Fundamentals

> You know, there may be some born leaders, but there are too few of them for us to count on them.
>
> —**Peter Drucker**

Vision

In this day and age, few of us have had the opportunity or need to sail a ship. I don't mean a sailboat or a waverunner, but an honest-to-goodness ship. Just over a century ago, ships were vital to the economy of any industrial nation. There was every kind of ship, from small rum runners to huge, hulking whaling ships that could spend years and years at sea.

Because none of us has ever had to pilot one of those kinds of ships, we have difficulty appreciating how difficult the task is. The ocean doesn't provide a sea captain with much in the way of landmarks, for example—that's why they are called "landmarks," not "seamarks." If the sky is overcast, telling which direction you are going is virtually impossible. There are no roads or markers to follow, but miles and miles of space in which to get lost.

Even worse, there are dozens of things that can potentially throw a ship off course, if not corrected for in advance. Tides and currents can move a ship very subtly off course; prevailing winds can help or hinder progress (or simply toss a ship about); even a poorly organized crew can turn a well-laid course into a major detour.

Why am I describing the perils of sea captaining in a book on project leadership? Because the parallels are really striking: the difficulties of successful project leadership are just as myriad as the hazards of navigating a ship, though few people realize this.

The most striking similarity, though, has to do with vision. Trading ships—and even the pirate ships that preyed on them—never set out without a specific destination in mind. A ship captain had to know precisely where he wanted to go, and have a clear plan for getting there (intact and on time, no less). They then used the "tools of the trade" to zero in on that destination: the compass, the sextant, up-to-date maps, and the ship captain's innate ability for dead reckoning—in other words, his ability to figure out the ship's location.

Without a clear understanding of where to go, a project leader is like a ship without a compass. A ship without a compass might have a destination, but the best that such a ship could hope to do is to wander around and pray that the destination lies somewhere ahead. All the while, that ship will be pushed around by the tide and the prevailing winds, wandering without direction.

Likewise, a project leader without a vision is likely to wander around aimlessly, becoming distracted by irrelevant things and working with the wrong people on the wrong problems. A project leader needs to be able to not only see the end result or goal, but also understand why the project is important, and why the goal must be met. A project leader should be able to tie his or her project back to the company's corporate strategies and describe that vision to the entire team.

With a vision, the everyday details of a project—motivating the team, making sure the project is on time, and moving each step in the right direction—become activities that the project leader does almost unconsciously to move the project forward to completion. However, if there is no vision, or if the vision is unclear or ambiguous, the project team will recognize this (perhaps unconsciously) and react poorly.

So what exactly is a vision, and how does one get a vision? A vision is simply the recognition of the true end point of a project. A vision is not just the last step of the project, or the result that is measured in the end, but also the overarching purpose for the project.

To understand vision, think of some successful athletes you know. Many professional sports teams as well as Olympic athletes use visualization techniques to make sure that they achieve their ultimate goals. These athletes not only practice on a regular basis, but also envision the final outcome. They mentally picture themselves swimming, running, or throwing faster and better than they ever have before, then envision themselves standing on the Olympic podium wearing the gold.

Athletes use this technique because it works: visualization has a powerful psychological impact. A good leader understands the power of vision and continually tries to create, communicate, and harness a vision within his or her own teams and projects. Of course, a vision might need to be clarified or modified as a project develops, but this is always done within the framework of goals laid down at the start.

Robert Greenleaf, father of the servant leadership movement, had these challenging words to say about vision:

> Why are liberating visions so rare?... [Half the problem] is that so few who have the gift for summarizing a vision, and the power to articulate it persuasively, have the urge and the courage to try. But there must be a place for servant-leaders with prophetic voices of great clarity who

will produce those liberating visions on which a caring, serving society depends (Spears and Lawrence, 2004, p. 3).

Visions must be communicated in order to be effective. Whether the vision comes from those above you in the corporate hierarchy, or whether the vision is one of your own design and imagination, a vision only stays in your imagination until you find a way to communicate that vision to your team.

Once communicated, however, a vision does several things at once. A vision

- gives team members a common goal to work toward, and common values to consider while working toward those goals.
- gives employees a framework for solving problems and removing bottlenecks.
- gives team members a system of metrics for success.
- opens channels for conflict resolution, allowing different people to assess methods, results, and attitudes.

Your first task as a project leader, then, is to find (or create) the vision and communicate that vision to your team. As the Reverend Theodore Hesburgh once said, "You have to have a vision. It's got to be a vision you articulate clearly and forcefully. You can't blow an uncertain trumpet."

Passion

During a presidential primary debate, the moderator decided to start with a "get to know you" question: he asked the candidates what their dream job was. The answers ranged from teacher to foreman in a steel mill (really?), but there was one candidate who candidly answered, "My dream job? I want to be President of the United States!"

That guy almost had my vote! No true leader can succeed without passion. Many people can manage a team to get things done, but without true passion for the activities you engage in, leadership seldom happens.

An individual can have good values, a spectacular dream, and a plan for realizing that dream, but without the burning passion to see the project through, that dream will ultimately just sit on the shelf as an unrealized idea.

Passion is the fuel that propels leaders forward. Whenever we meet a true leader, we are often astounded at the energy and determination these individuals have for their specific cause. That energy and determination are what break down barriers and provide the fuel for overcoming challenges.

Passion goes beyond being dramatic or emotional—passion is not merely enthusiasm. Rather, passion is an unfailing commitment to a vision. Whereas enthusiasm starts strong but quickly fades in the face of time and challenges, passion only increases in intensity and duration during tough times.

A strong leader shows passion by reaffirming the core values of their organization and by communicating their vision with excitement. They "stoke the inner fires" and use that energy to deliver service and value. As a result, their passion often becomes contagious. Their passion can be used to inspire large groups of people, but, more importantly, that passion can be used to inspire small groups of people to extraordinary results.

Any veteran project leader will tell you that inciting people's passions can be difficult, especially for very mundane or repetitive projects. But a good project leader knows that passion is not just about enthusiasm for the project (although this is key).

A good project leader is often just as passionate about the processes, people, and purpose underlying a project as he or she is about the project itself. They love learning about and testing new methods. They take an interest not only in individuals and their abilities, but also in the success and development of those individuals, and they do their best to understand, communicate, and truly live the vision.

Once you have developed this kind of passion, work becomes more fun and teams transform into high-performance machines.

Working through a Team

Your time as a project manager will not be remembered by what you accomplish, but rather by what you are able to lead your team to achieve. Your team's results will be the hallmarks of your success as a leader.

Why is that? On the news, in trade magazines, and in history books, we learn about true leaders and their qualities—some are even given an almost heroic status. So why focus on team accomplishment here, rather than on leadership qualities?

There are hundreds of books claiming to be able to describe, in exact detail, what leadership is and how to achieve leadership roles, but leadership is a quality that is hard to measure. I doubt that real leadership can be captured in a book, much less a bulleted list. Nonetheless, I will share with you one definition that I think comes close:

A leader is an individual who inspires, cajoles, encourages, threatens, cheerleads, and serves a group of people to get a specific task, or series of tasks, done.

When leaders have done their job, the people they led should be able to say, "We did it ourselves!"—they will feel like they accomplished everything themselves, rather than having been led to do anything.

The Fundamentals

VISION

Without vision, the end is never in sight.

PASSION

Without passion all the rest is lost.

WORKING THROUGH A TEAM

A leader is an individual who inspires, cajoles, encourages, threatens, cheerleads, and serves a group of people to get a specific task, or series of tasks, done so that at the end of the project, the group can say, "We did this project ourselves!"

You've probably recognized by now that this is a very different leadership style than is practiced in most organizations. I find this surprising, since this kind of leadership is exceedingly effective in leading project teams effectively to their ultimate goals.

When a project leader is able to serve his or her team, the team as a whole becomes better able to deliver on the desired results.

Although there is much wisdom in the popular lore on leadership, what I am suggesting here is a different kind of leader, the sort of leader that is down "in the trenches" serving his or her team.

This kind of leader is knee-deep in the workings (and the problems!) of the team, inspiring them as they go and handing over the credit to the team, instead of working for personal kudos and self-congratulation.

If you meet your people's needs, you will find that they will get you everything you need in return. If you make your team look great, you'll look great.

Work through your team to accomplish the best results.

Chapter 1 Review

To be successful, project leaders need to develop critical leadership fundamentals: vision, passion, and the ability to work through a team.

Vision

A project leader must have a clear understanding of the team's direction and goals. A vision is the recognition of what the successful outcome of a project looks like. Without a clear vision to lead the team toward, the project leader and his or her team are likely to be distracted by irrelevant tasks and focus on the wrong activities. Project leaders must understand why the project is important and how it fits within the organization's bigger goals. The project leader must also be able to share this vision with his team and encourage the team to embrace the vision as well.

A vision

- unites the team under a common goal.
- provides a framework for problem solving.
- creates a system of metrics by which to measure success.
- allows for optimal conflict resolution because the team is working toward a common goal while abiding by like values and principles.

Passion

Passion is crucial to leadership success; in fact, without passion a leader would be unable to succeed. Passion is the fuel that drives the team forward toward its vision. Passion breaks down barriers, overcomes obstacles, and allows a team to stay focused on its goals even when times are tough. The project leader must be passionate about the project, the people working on the project, and even the underlying purpose of the project itself.

Working through a Team

A project leader's tenure is not marked by what the leader accomplishes, but what he is able to lead his team to accomplish. No project leader is capable of completing every step of a project on his own; the project leader is absolutely dependent on his team to keep the project moving forward. In order to bring the project to a successful conclusion, the project leader must be able to work through his team to inspire them to work passionately and embrace the team's vision. When a project leader works through his team, the team will have the sense that they accomplished the project themselves.

2

The Project Manager as Entrepreneur

> Entrepreneurs have a mindset that sees the possibilities rather than the problems created by change.
>
> **—J. Gregory Dees**

Early in my life I had the good fortune of working with a small publishing company in Boston that specialized in "consumer content" for the health care industry. The president and founder of the company, a large, loud man named Terri, was perhaps one of the craziest—and one of the most successful—company presidents I have known.

Among Terri's numerous odd habits, he would walk the halls of the company, pop in on his team leaders, and say in a booming voice, "Hi there! How have you made me money today?"

Each of Terri's new team leaders became visibly flustered the first few times that he did this. In fact, I once overheard a new team leader complain, "I can't believe he asked me that—I'm doing my job, aren't I?"

The point that Terri was trying to make was that the role of team leader in his organization was not just to "do a job." He wanted each team leader to focus on the goal of making money for the company—whether that meant providing a new product, improving an old product, getting new customers, or simply increasing efficiency.

Terri set up his company with the view that the team leaders were small business owners themselves, not just managers or directors. Once a new team leader got this core idea, their teams usually flourished. The responsibility was squarely on the shoulders of the team leaders to improve their products, develop a rapport with customers, and learn to be sensitive to the needs of customers and employees alike.

I heard Terri's philosophy summed up nicely by the head of a major project management office—a woman considered one of the premier experts in her field. During a local symposium, she made the following remark:

> A project manager should be an entrepreneur.

This one line encapsulates the main message of this chapter. Project managers are not just managers; they are also entrepreneurs. As the entrepreneurs of a small team, project managers must own their projects. They must understand where the business is going, and where the business falls within its industry and the overall market. When project managers understand not only how the project works but also how the project works within the corporation's bigger picture, they can help their teams succeed.

If you understand these principles, you are already on your way to increased efficiency and successful goal attainment when you lead a project team.

The Project Manager as Entrepreneur

Take a few moments and reflect on the responsibilities of a successful entrepreneur. Entrepreneurs must deal with a variety of people: for example, customers, personnel, vendors, and financial backers. They must develop and follow a budget. They must deal with the entire business process, from marketing to delivery, all the while overcoming market forces and unknown factors, and ultimately steer their projects to completion.

As organizations become leaner, project managers are asked to lead their projects in a more dynamic way. A project manager must be passionate and involved enough to inspire an entire team to get behind the endeavor and work toward a successful conclusion. While working through this process, the project manager must manage internal and external stakeholders, finances, quality customer service, and the risk and implications thereof.

In many ways, a large project can be like a business. Projects redefine markets and change the course of business on a regular basis. This entrepreneurial role requires the project manager to move away from dealing with the standard forms and documents and into the bigger task of running the equivalent of a company.

Within the larger organizational structures, this is nothing new. Many project managers run projects that are the size of Fortune 500 companies. The difference is that these project managers do not use management techniques but instead rely more on leadership. This shift is necessary because the projects are so large that they cannot deal with all of the workers on an individual basis and must step into a leadership role to control and execute a project of this size.

Most importantly, an entrepreneur must own his or her project. There is no such thing as punching out early, or deferring responsibility, or trying to get away with half a job. Successful entrepreneurs understand that they are responsible for their own actions—and for the individuals and resources entrusted to their care as well.

A project manager has a similar role. Project managers must own their projects in just the way that an entrepreneur would. Successful managers take responsibility for their actions, and for the individuals and resources entrusted to their care.

Steering the Project as an Entrepreneur

The first tasks of an entrepreneur are to develop a business plan and to put processes in place that get their goods and services into the hands of the customer.

A project manager usually has a more limited role than this: the business plan is provided by the higher-up executives, and some of the processes might already be put into place. Although the project manager might not compose the business plan, he or she must create a document called a "project charter." The project charter is a starting point that defines the project, identifies the project's status, and explains how the project will be completed. This is an important document for a project manager, considering that the project charter is derived from the original business plan.

Although the business plan may not be created by the project manager, planning how the project is going to get done on time and within budget, and who's going to do the work, is certainly something that both entrepreneurs and project managers must determine. Both must also consider and understand the project's financial implications and how the project fits in with the organization's and the industry's bigger picture. They cannot be myopic and worry only about the project or business itself.

Key Points

- The project manager must become an expert on their project.
- The project manager must be able to stay organized and keep the desired results in perspective.
- The project manager must be able to communicate the "big picture" to their team—without micromanaging every task.
- The project manager must understand where their efforts will be put to best use.

Like entrepreneurs, project managers must become experts on their project. They need to know everything about the project from the grand vision to the minutest details. They need to be able to visualize each step from the project's conception to the summary report given to the higher-ups upon the project's completion.

The best entrepreneurs are able to stay organized and keep the end in sight. This is an important characteristic of project managers as well. Even though managers need to be experts on the details of their projects, they must also be able to keep the big picture in mind. They also need to communicate that big picture to their subordinates and monitor their progress—without interfering with their work.

Perhaps most counterintuitively, entrepreneurs must know when to step away from what they are doing. Even though entrepreneurs are expected to know everything about their business, their talents do not lay in running its day-to-day details (if they did, they would be handling those details for someone else). In businesses that fail, employees often report being overwhelmed by the amount of micromanagement that goes on from the top; resisting the temptation to micromanage is a key skill of a successful project manager.

Dealing with People as an Entrepreneur

A project manager never accomplishes a project alone. Teamwork is crucially important—so important that there are two separate chapters dealing with the topic in this book. But consider for a moment the other sorts of relationships that the entrepreneur must build in the course of doing business.

James Hunter, author of *The Servant*, once said that business was nothing more than a series of relationships—and that the most successful leaders had learned to develop healthy relationships with their CEOS. Hunter wasn't referring to the firm's chief executive officer, but rather the firm's Customers, Employees, Owners, and Significant Others.

Key Points

Successful entrepreneurs understand the importance of their relationships with other people; relationships are the foundations of any successful business or project.

Project managers must take care to develop healthy relationships with their CEOS:

- Customers
- Employees
- Owners
- Significant Others

Let's continue with our analogy of the project manager as entrepreneur as we explore the leader's relationship with his or her CEOS.

Customers

Entrepreneurs must be able to attract, listen to, serve, and leverage their customers. They work on solving customer problems and maintaining relationships in order to guarantee future business. Customers expect a quality product, on time, at a reasonable price. If something in the process goes wrong, the entrepreneur must work with the customer to resolve the issue.

Project managers must deal not only with the firm's external clients, but also with its internal clients. The internal clients are, of course, individuals within the organization—the people doing the work. Internal clients have expectations as well: to be treated with respect, to be given fair compensation, and so on. Some project managers don't view the people within the firm as customers and thus treat them with less respect; however, without these "internal clients," the organization wouldn't have a chance of meeting the needs of the external customers.

Entrepreneurs serve their customers; project managers must do the same, and consider customers both inside and outside the organization.

Employees

Perhaps the strongest analogy with the project manager comes from the way that an entrepreneur must deal with his or her employees. Although entrepreneurs can designate authority for an aspect of their business, they do not delegate responsibility.

Ultimate responsibility for the success of the business (and the happiness of the customers) always lies with the entrepreneur. Their task is to empower employees and provide them with the tools and guidance necessary to do the job, so that success is guaranteed.

The only way to truly empower employees is to serve them by working to meet all of their needs and removing as many obstacles to a successful end result as possible.

In businesses that fail, the common pattern is that the manager does not give enough authority to his or her employees, and so ends up having to "put out all the fires" on his or her own. This eats away at the time needed to actually work on the project.

Owners

Owners can refer to stockholders, taxpayers in the community, or investors. Owners need to get a fair return on their investment or the firm will not be in business for long. Yes, entrepreneurs may be the sole owner of their firm; however, even the most dedicated entrepreneur will not be in business for long without fair returns for his or her time, effort, and financial investments into the firm.

Project managers have another owner to serve: internal business owners, also known as "sponsors" and "executive sponsors." The sponsor is the individual heading up the project on the business side. A sponsor is not assigned to the project, but rather is the person who assigned the project, and another type of internal customer. An executive sponsor, on the other hand, is the person who is paying for the project if the sponsor does not hold the purse strings.

Significant Others

Besides their customers, entrepreneurs must work closely with vendors and suppliers as well. They must develop healthy working relationships with these significant others to guarantee prompt delivery and fair prices. And they must balance loyalty with practicality in their bids.

Project managers must also deal with stakeholders. These are individuals who are significant to the project, but may have no direct influence on the project. For example, if the purpose of the project is to compile a new customer service database, the customer service representatives inputting the data may not have much say in the project but could make or break your success with the project. Therefore, we'd consider them stakeholders.

For the entrepreneur and the project manager, long-term leadership success rises and falls with the individual's ability to maintain healthy relationships with the CEOS. An entrepreneur or project manager may succeed temporarily without developing solid relationships, but the leader's ability to influence, inspire, and guide people and meet their CEOS' needs will determine the fate of the project or business.

Dealing with the Unknown as an Entrepreneur

Even the best-planned project, headed by the most competent leadership, will face some unknowns—accidents, mistakes, new market conditions, changing goals, shifts in customer demands, supply quandaries, changing deadlines, new research...the list goes on. Every one of the unknowns can be, and has been, used as an excuse for the poor performance of a team.

An entrepreneur knows that there are no excuses, no reasons for failure to point out to the higher brass. If a business fails, the business fails—period. The job of the entrepreneur is to try to avoid failure at all costs—even in the face of the unknown.

A good project manager can plan ahead for some of these unknowns. They can build in back-up systems for employees and processes, pad schedules and deadlines, and try their best to stay informed. Ultimately, though, project managers simply have to accept the fact that unknowns will occur. They must be nimble, receptive to new information, and decisive in the actions they take to solve problems.

In order to move forward with decisive actions, they must have sufficient information to make the right choice, and be able to take responsibility for the decision both financially and mentally.

In Chapter 7, we'll discuss how project managers can cope with changes in the project plan and teach their teams how to react to unexpected obstacles.

The Failure to Be an Entrepreneur

We all know small business owners who have succeeded, and we all know small business owners who have failed in their endeavors. What conclusions can we draw from that knowledge? How can so many entrepreneurs fail—and, by the same token, how can so many succeed?

When a project fails—and from time to time, they do fail—your average project manager will point up or down for reasons why his or her project is not going well. They will blame the upper-level leadership that conceived of the project, the administration that failed to give them the resources that they needed, and the employees whose work was subpar...but never themselves. (By the way, this is the number-one sign that someone has done little work as well.)

Once a project has failed, the best thing a project manager can do is stop, ask him- or herself, "What could we do better next time?" and learn from the mistakes. In project management parlay, this process is called "lessons learned." Many entrepreneurs fail or go bankrupt at least once before achieving major success. What enables entrepreneurs to ultimately achieve that success is their ability to learn from their mistakes. Therefore, being able to summarize the lessons learned when faced with failure can be the deciding factor in future success.

In the final analysis, project managers are responsible for their project, just like entrepreneurs are responsible for their business. Sometimes entrepreneurs succeed, and sometimes they do not. Likewise, there are times when projects succeed, and times when they do not. But the key factor in the success or failure of a project is what the project manager is willing to do to get the project done.

Some projects are relatively easy, but the ones that are worthwhile take time, ingenuity, skill, and—dare I say—wisdom. Some projects will require working those late hours. Some will require extra research, or renegotiating with suppliers, or speaking with clients every day. A project manager needs to see him- or herself as the head of a small business, dealing with all the people and problems that this entails.

If a team project is a small business, then the corporation as a whole can be seen as a regulating market. This market has multiple demands and requirements, which must be understood, navigated, and used to make a project work. If the project manager can understand this analogy, he or she can begin to see how the "market" or the corporation works and begin to work with those market forces, rather than against them.

Something amazing happens when a blossoming leader achieves this understanding. A project manager who gets this message will make tremendous strides and become a highly sought-after individual within the organization. This is so because such a manager will routinely deliver results and develop a knack for working with others.

People will routinely want to work with such a manager on projects, and that loyalty will continue to pay dividends on each new project.

Getting the Entrepreneurial Mind-Set

Remember how Terri used to roam the halls of his publishing company, asking his team leaders how they had made him money that day? As brazen as this habit sounds, I am convinced that Terri would not have asked the question if he did not think that his team leaders were capable of changing their mind-sets.

His question was meant to dislodge familiar ideas of management and get people thinking of their teams as small businesses, with the team leader as entrepreneur. And the strategy seemed to work—each new team leader eventually understood Terri's strategy and was able to transform him- or herself into a small business leader.

How did they make this transition? By adopting the sets of traits, habits, and attitudes of an entrepreneur. Terri's prodding encouraged team leaders to assume the characteristics of an entrepreneur, and model their actions and thought processes accordingly.

Consider the mind-set of the successful entrepreneur. What are the defining sets of traits, attitudes, and habits that we see almost exclusively in entrepreneurs (and that others—especially failed entrepreneurs—lack)?

Traits

An entrepreneur is

- hardworking.
- honest.
- friendly.
- a storyteller.
- curious.
- resourceful.
- visionary.
- a motivator.

Habits

An entrepreneur

- works on a business, not in a business.
- keeps the final goal in sight.
- plans ahead.
- puts customers first.
- works on his or her basic skills.
- tries to stay informed.
- remains flexible.
- knows how to use people's strengths.
- has a talent for overcoming people's weaknesses.

Attitudes

An entrepreneur has

- a positive attitude.
- faith in his or her project and abilities.
- an ability to see the best in people.
- contagious enthusiasm.
- a taste for success.
- a love of his or her product.

For some strategies on how to use this within your team, please go to our companion website, www.ManagementToLeadership.com.

Every motivational speaker, management expert, and guru has their own list of leadership qualities, but there is very little out there about actually developing these qualities. These "experts" appear to assume that there are little switches inside us labeled "positive attitude," "contagious enthusiasm," and so on, and we can just flip these switches on at whim.

So as not to fall into the same trap, I have provided seven tips for actually developing the entrepreneurial qualities that will make a difference to your project:

- Think concrete.
- Set SMART goals.
- Put your thoughts on paper.
- Start small and make a habit.

- Role-play.
- Look to role models.
- Set aside time for personal development.

These tips have been gleaned from actually watching people as they made the transition—and as they learned to answer Terri's all-important question, "How have you made me money today?"

Think Concrete

Attitudes, virtues, and strategies are all very abstract. Abstraction is not, by itself, a bad thing. Human beings have become successful partly through their ability to summarize large amounts of data, analyze patterns, and talk about them in very concise, abstract terms. But this ability can also be misleading when the abstract term takes on a life of its own—without clearly specifying the precise actions that make up the abstract attitude or virtue.

For every trait you want to develop, think of at least five concrete actions that best demonstrate that attitude. These can be things you do on a regular basis, or ways that you would respond to specific situations as they arise. For instance, if you believe that you need to be more resourceful, what are five concrete actions you could take to develop this trait?

Visualize yourself performing these actions. "Listen" to what you would say, and "watch" what you would do. Fill in the details of the context: where you are, who is with you, and even what you are wearing. The more you think about these concrete actions, the more you will prompt yourself to do them regularly.

Of course, visualizing isn't a fair substitute for action. Once you've visualized yourself completing these actions, you must begin actually performing them.

Set SMART Goals

One of the best ways to motivate yourself to take action on your traits is to set goals for yourself. The best goals are SMART goals:

Succinct	Goals should be clearly and concisely defined to avoid ambiguity.
Measurable	Goals should have "metrics," or benchmarks that identify progress.
Attainable	Goals should be achievable, but require that the person stretch his or her limits.
Realistic	Goals should be sensible and have a realistic outcome.
Time bound	Goals should have a timeline; a goal without a deadline is just a dream.

Visualization is a necessary first step, but only action will help you achieve your desired outcomes. Setting SMART goals will help you make progress toward achieving otherwise abstract attitudes, virtues, and strategies, and enable you to measure your specific progress toward becoming the leader you want to be.

Put Your Thoughts on Paper

Once you've set SMART goals for yourself, post reminders of your aspirations around the office. Most great people think on paper. There is a psychological interaction between being able to write down your goals and seeing them on paper; this develops a feedback loop in your brain and jumpstarts the associative nature of your brain.

Recording your goals and ideas on paper gives your mind the ability to make your visions into reality. Writing the words also helps reinforce and reemphasize the idea, and allows your brain to stop making continual efforts to remember the goal.

I once came across an old study about successful people. The researchers selected a random grouping of college students, interviewed these individuals after they graduated, and then tracked their progress through their professional careers. The study found that the graduates who set goals and wrote down those goals achieved substantially higher levels of success than their counterparts who failed to set goals for themselves at all, and higher levels of success than those who set goals but did not write their aspirations on paper.

Start Small and Make a Habit

Don't start out by trying to develop a complete mind-set, or a whole new attitude. Start small. Perform actions that are so easy that they seem like little at all. Practice these new actions every day. The common wisdom is that you must practice an action for twenty-one days before the action becomes a habit. If you find that maintaining the action for twenty-one days is easy, then make the task a little more difficult and try for another twenty-one days.

If you cannot maintain the action for twenty-one days, set a slightly less ambitious goal. The smaller the steps you take, the less likely you are to invent excuses and avoid doing them.

Role-Play

Odd as this tip might sound, pretending to be the person you want to be is a good first step to actually becoming that person. Building an identity for ourselves is always a difficult process, but if you immerse yourself in the "role," you will find it easier to perform those actions you need to perform.

For example, you might not be a very cheery person, and not prone to smiling at random. But perhaps you want to develop a cheerier demeanor. Well, role-play as if you were a cheery manager. Force yourself to smile, even if it feels like an act. Think of the lines that would be delivered by an actor playing the part of "manager with a positive attitude."

Role-playing in this way will feel a little forced at first, but keep trying. You brain does not make a distinction between habits you develop genuinely and habits you develop while role-playing. And so playing the role of the manager you want to be can be a great way to "trick" your brain into developing positive habits.

In the appendix of this book, you'll find a section titled "Leadership Exercise: Role-Playing the Medieval Court." Read through each of these different project management roles, and then select the role that you feel least comfortable with.

Focus on assuming your selected role until the character seems natural to you. At that point, you may select another role to practice. Continue until you've familiarized yourself with every position in the medieval court.

Look to Role Models

Don't feel like you have to break new ground and discover how to be an entrepreneur from scratch. Look to others who exemplify the attitudes, virtues, and habits you want to develop. Talk with them, watch them work, and use them as examples. You can even use them to modify the "role" you act out as you role-play your position.

Set Aside Time for Personal Development

If you really wanted to, you could spend an entire day on these processes—and on yourself. Of course, nobody has that kind of time. But you should set aside a little chunk of time each day to work on your mind-set explicitly.

Key Points

Adopting the mindset of a successful entrepreneur is not as simple as flipping on an inner switch. Following these techniques will help you adopt otherwise ambiguous traits, habits, and attitudes:

- Think Concrete
- Set SMART Goals
- Put Your Thoughts on Paper
- Start Small and Make a Habit
- Role-play
- Look to Role Models
- Set Aside Time for Personal Development

Do this in a quiet place, without any distractions, and feel free to write down any results, problems, or musings. Then spend the rest of your day working on your project.

Consistency is more important than the total amount of time spent when it comes to developing mind-sets. If you find yourself crunched for time, try multitasking. For example, you might use audio CDs to keep up with important reading while you commute to and from work each day.

Personal development goes beyond just working on your mind-set. You should also work to continuously increase your knowledge of your industry and its place within the market.

I was lucky enough to watch several team leaders use just these techniques to change their mind-set from that of a manager to that of an entrepreneur. The results made Terri a great deal of wealth, to say the very least. Try some of these techniques for yourself and see what works.

By working to constantly learn more and improve your skills, you will achieve a better understanding of the entire environment that surrounds you: your projects and their place within the organization, the industry, and the entire market.

If you do not yet understand how to be an entrepreneur within your own organization, you need not look far to find useful tips and advice. Any good book on being your own boss can give the project manager tremendous hints on how to become an entrepreneurial project manager.

Chapter 2 Review

Project leaders are more than mere managers. A project leader's role is most similar to that of an entrepreneur. Entrepreneurs own their businesses. They know the direction in which their company is headed and how their company fits into the industry. Entrepreneurs can't just clock out or take the day off because they know that their company depends upon them. Similarly, project leaders cannot defer their responsibility because they must own their projects. Project leaders need to understand where the project is headed and how the project fits within the organization's larger goals.

Working with CEOS

A project leader must cater to the CEOS of a project:

- Customers
- Employees
- Owners
- Significant others

The project leader's relationship with the projects' CEOS is critical to a successful outcome. Each stakeholder has unique needs and expectations.

Key Points

Like an entrepreneur, a project leader must be all of the following:

- The expert on the project
- Organized and able to keep the team moving forward toward a successful outcome
- Capable of communicating the "big picture" and promoting the team's vision without becoming a micromanager and looking over team members' shoulders
- Able to use the team's resources (including human resources) where they will be most effective
- Comfortable working with the project's CEOS
- Able to deal with the "unknowns"
- Prepared to deal with potential project failures and able to learn from his or her mistakes

Project leaders can develop entrepreneurial skills through practice and perseverance. Learning to think concrete, setting SMART goals, developing smart habits, role-playing, and setting aside time for personal development are key to learning to think like an entrepreneur.

3

Understanding Teamwork

The era of the rugged individual is giving way to the era of the team player. Everyone is needed, but no one is necessary.

—Bruce Coslet

In most organizations, a "team" is defined as a group of people assembled for a specific purpose, usually to complete a given project or task. Anytime one of these groups is formed, there is an inherent expectation among the organization's leadership that there will be teamwork between the members, yet merely assigning a group of individuals who work for the same organization to the same group does not create a team.

One of the first things a new project manager often realizes is that he or she is incapable of completing a project alone. The project manager must depend heavily, and sometimes exclusively, on the team that has been assembled. Unfortunately, not every team is capable of teamwork.

The Fallacy of Teamwork

Industries are becoming more competitive each day. Globalization of the world's economy and electronic media such as the Internet create a "flat" surface for all organizations, and make finding strategic differentiators between these firms more challenging.

Superior teamwork is one way to level a strategic advantage of an organization; however, most organizations do not invest the time and effort in order to develop this teamwork.

The word "teamwork" is overused in the business world. There are thousands of organizations in the world that claim to possess great teamwork simply because their organizations are filled to the brim with teams. I've been told by corporate executives, "We all work in teams here," only to discover later that the teams are moved around every four months. This type of constant shifting breaks up the dynamics of a group, destroying their chances of achieving teamwork. These corporations are just interrupting their teams during the stages of team development without actually gaining any of the advantages of genuine teamwork.

TEAMWORK

n.

1. Work done by several associates with each doing a part but subordinating personal prominence to the efficiency of the whole.

Teamwork is the ability to work together toward a common vision. The ability to direct individual accomplishments toward organizational objectives. It is the fuel that allows common people to attain uncommon results.

Andrew Carnegie

There is a pervasive fallacy among corporate leaders that creating teamwork is as simple as putting together a random grouping of individuals, giving them an assignment, and having them work on the project together. In some circumstances teamwork can happen this way, but most the time when a group of individuals is working together, they are not exhibiting real teamwork—just coexisting peacefully. This environment of artificial harmony is one of the first clues that something is awry.

A second fallacy of teamwork is the assumption that individuals can work as a team from the very beginning. This is not true. In most situations, teamwork is something that has to be developed over time.

Teamwork cannot be assigned or dictated by organizational structure. Certainly, organizational structure can encourage or discourage teamwork, but cultivating an environment conducive to true teamwork requires a tremendous amount of time, effort, and energy. In most large corporations, the time necessary to develop true interdisciplinary teamwork is not given.

Redefining How the Team Works Together

Although achieving an environment conducive to teamwork is not as simple as assembling a group of workers and labeling the group a "team," teamwork is a powerful tool that can be cultivated within organizations of any size.

Most organizations already know that teamwork is a good thing. They want to use teams to their advantage, but they mistakenly believe that if senior management says, "You're going to work together as a team," to a group of individuals, teamwork will magically occur.

These same organizations are often so shortsighted that they are unwilling to expend the time and effort to develop exceptional dynamics within their teams, and instead settle for individuals merely showing up and halfheartedly working with one another.

Developing a good team requires time, energy, and encouragement. Teamwork is only possible when every team member understands the following:

- How the other team members work
- How the other team members think
- How each member of the team can proactively support and encourage his or her teammates
- How each team member's role will aid in achieving the project's desired results

Most groups of employees can work together, but lack the necessary commitment and trust to truly achieve teamwork. Members of true team should not only understand their individual jobs and roles within the team, but also have commitment to and a sense of accountability for the entire organization. They should support each other. They should work together like a well-oiled machine, working proactively to achieve the expected goals in the specified time.

I once worked with a remarkable team in Boca Raton, Florida. We had worked on similar projects three or four times in the past and had become quite familiar with one another, to the point where we could switch roles very quickly.

During a major client crisis, I was stuck in an airport while attempting to work through an issue. Luckily, the entire team was there. The technical person took over working with the customer. One of the quality assurance (QA) people took over the technical work. Because we understood one another's roles so well, we were able to navigate through a major issue easily by sharing the workload.

The customer was astounded at how quickly and efficiently our team had solved the problem. The client had estimated that the issue would require two to three weeks to resolve—we handled the task in a day and a half.

In the face of a major crisis, no one stood back and said, "That's not my job," or pointed fingers. Because the entire team shared a steadfast commitment to the organization and knew that every member of the team was accountable for the final results, we were able to pull together and save the day.

Defining a High-Performing Team

True teamwork is much bigger than an organizational chart or random groupings of individuals.

True teamwork is uncomfortable at times, but the rewards are well worth the step outside of your comfort zone.

True teamwork can be a strategic differentiator between you and your competitors.

Many people talk about teamwork and "high-performing teams," but fail to define the attributes of these teams or provide criteria by which to judge these teams.

A high-performing team has the following traits:

1. Trusts and has confidence in the other team members
2. Clearly understands what the team must do
3. Embodies the key mission within their everyday lives
4. Takes responsibility for the team's long- and short-term goals
5. Displays loyalty, enthusiasm, and zeal for their mission
6. Encourages differing views and perspectives to gain an overall understanding
7. Develops a consensus on strategies, tactics, and resources
8. Defines roles clearly
9. Cross-trains so that team members can switch roles quickly when necessary
10. Encourages risk taking and experimentation for new solutions
11. Allows constructive criticism and helpful feedback to improve team members
12. Uses errors to improve the process—not to rebuke other team members
13. Develops a verbal or written history
14. Rotates heroes throughout the team
15. Has leadership that is clear, strong, and able to adapt to changing situations quickly

Trusts and Has Confidence in the Other Team Members

A high-performing team has worked together long enough to understand that they work better as a team than as individuals, and they know that the entire team is committed to the project.

Clearly Understands What the Team Must Do

Members of a high-performing team know that it is vital that all members understand the team mission as well as what tasks each individual member must complete. If the mission or the breakdown of assignments is unclear, the members will delve into greater detail to ensure that everyone understands both the specifics and broader themes of the team's mission.

Embodies the Key Mission within Their Everyday Lives

High-performing teams have so internalized their mission that team members reflect the team's mission statement in their actions, words, and deeds, both in and out of the office. Members of high-performing teams find ways to incorporate the team's mission into their everyday tasks and attitudes, not just into team-oriented activities and assignments.

Takes Responsibility for the Team's Long- and Short-Term Goals

Every member of a high-performing team assumes responsibility for achieving the team's goals. Members are able to both enunciate and internalize goals, and are therefore able to identify new opportunities to foster the broader, long-term objectives through short-term successes and task-oriented work. High-performing teams recognize that if every player is thinking about and working toward the team's objectives, new and inventive ideas will present themselves to all, rather than expecting one individual to push the team toward reaching said goals.

Displays Loyalty, Enthusiasm, and Zeal for Their Mission

High-performing teams display enthusiasm for and incredible interest in their mission and its outcomes. This enthusiasm goes beyond the standard parroting of the corporate philosophy when the organization's higher-ups are within earshot. Members of these teams internalize the mission, and the comprehension of the team's mission moves from an intellectual understanding of a target to a physical, visceral need to achieve the desired outcome.

Encourages Differing Views and Perspectives to Gain an Overall Understanding

Members of a high-performing team recognize that differing viewpoints encourage broader thinking and foster better long-term solutions. Members are not afraid to bring up new ideas and opportunities that could improve the team's efforts, even though these ideas might not be part of the team's specific goals.

Develops a Consensus on Strategies, Tactics, and Resources

High-performing teams develop a means of consensus building in order to ensure the inclusion of all team members in the planning of strategies and tactics. Once everyone has expressed an opinion, the leader of the team must establish a consensus among team members to deliver the best results. All individuals then stand behind the decision as a team, regardless of their previous position on the issue.

Defines Roles Clearly

For a high-performing team, the clear delineation of roles is vital. Roles must be assigned and defined with consideration to each individual's level of knowledge and expertise. The team recognizes that there might be individuals with specific talents that fall into natural roles, but all team members are encouraged to provide ideas and feedback for the betterment of the team, even if that duty doesn't fall within their assigned role.

Cross-Trains So That Team Members Can Switch Roles Quickly When Necessary

Even though roles are clearly defined within high-performing teams, members of these teams cross-train on different roles. This ensures that if one team member is unavailable, other team members can step in to fulfill that role.

Encourages Risk Taking and Experimentation for New Solutions

High-performing teams are typically given more challenging assignments because the team is known to be dynamic, capable of making decisions quickly, and able to complete work faster than other teams. Members of these teams are unafraid to take greater risks, try new ideas, and experiment with different ways of doing things because they know that their team will back them up and support them.

Allows Constructive Criticism and Helpful Feedback to Improve Team Members

High-performing teams have difficult people and problems to handle on a regular basis. Members of these teams are confident enough to welcome constructive criticism of their behavior and abilities in challenging situations. This feedback is focused on helping teammates develop better strategies and identifying areas that can be improved, as opposed to commentary on personal issues. Constructive criticism, when offered correctly, helps maintain an atmosphere of trust within high-performing teams.

Uses Errors to Improve the Process—Not to Rebuke Other Team Members

High-performing teams use mistakes to increase their level of effectiveness. These teams do not place blame for issues on the individual, but take responsibility for the error as a team. This approach allows the team to learn from the mistake and avoid similar issues in the future.

Develops a Verbal or Written History

High-performing teams continually reinforce their core ideals. One of the most effective ways to do this is to develop written and oral stories that continue to teach and reinforce the team's ideal. These narrative histories incorporate the team's general strategies, mission, and long-term goals. Finally, these histories become a teaching tool for future team members, who will replace those original members that inevitably move on to other teams or organizations.

Rotates Heroes Throughout the Team

High-performing teams are constantly on the lookout for ways to recognize individual team members for a job well done. The team makes an effort to regularly identify and praise the team's "hero" for saving the day. Because the team is constantly making an effort to recognize quality work, the team's "hero" label will rotate regularly, rather than continuously falling to one individual.

Has Leadership That Is Clear, Strong, and Able to Adapt to Changing Situations Quickly

The leader of a high-performing team is highly adaptive and able to support the team in every endeavor. This leader must be confident in his or her abilities, and able to integrate, interact, and coach members of the team.

The leader should be able to identify potential problems before they arise in order to prepare the team to move on through difficult situations and generate solutions. The leader encourages and serves the team to make sure that each team member feels heard and supported.

For some techniques and tools that will help start the discussion of a high-performing team, please see our companion website, www.ManagementToLeadership.com.

Strengthening the Team Triangle

In order for a team to be classified as "high performing," there must be efficiency on all sides of the team triangle (see Figure 3.1). A team cannot be defined solely in terms of its leader, members, or task. Rather, teams are composed of a conglomeration of three elements: every team must excel at the task, team, and individual levels (the three sides of the triangle) to accelerate its performance. The team, composed of individual members and led by a project manager, must approach each task with the balance of control and autonomy required for effective results.

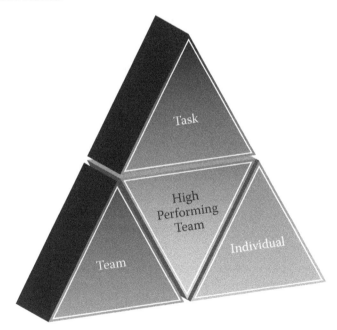

FIGURE 3.1
High performing triangle.

Each area of the team triangle must meet different standards to be effective, but every part of the triangle is of equal importance and requires the same level of attention from the project leader. A good friend of mine, Bobby Robinson, compares the team triangle to a three-legged stool. If one side of the stool is weaker or shorter than the rest, the stool will be nearly impossible to sit upon; however, if the three sides are equal in strength and size, the stool creates a triangle, the most stable structure possible.

Let's explore each side of the team triangle in greater detail.

Tasks

The task level of the team triangle requires that every individual in the team understands two important points from the outset:

- What is the expected outcome of the task?
- What are the guidelines for completing the task?

It is crucial that team members understand not only the expected results, but also the guidelines—how they are expected to execute the task in terms of quality level, ethical considerations, and project priorities. Together, the defined expectations and guidelines should paint a clear picture of what the task looks like and demonstrate how the task fits within the project as a whole. Team members should be able to define how the task corresponds with other tasks and how the individual activity contributes to the success of the project.

Once the outcome and guidelines of the task are clearly defined, allowing team members to understand the big picture, individuals must then identify and understand the process of getting from the starting point to the finish line. What needs to be done now, next, and later to complete the activity? What happens upon completion of this task? How does the team move to the next activity? Each of these questions must be answered to ensure the task achieves its goals. Team members should understand their specific roles and responsibilities within the given task, including the expectations for quality requirements. Members should also know what resources and technology they have available to leverage the effectiveness of the task from the beginning.

Individual

As the engine that drives a project toward completion, individual team members and the team leader are critical to the success of the project team as a whole. To achieve the highest stage of performance, four primary aspects of the individual must be considered:

- Commitment
- Energy
- Orientation toward results
- Autonomy

Commitment

First, what is the individual level of commitment? How devoted are the team members and team leader to getting the job done? This level of commitment will be reflected in the team's ability to complete tasks successfully and in the quality of the finished product (or in the final outcome).

Project leaders must take steps to build commitment at the individual level to achieve the best results. Involving individuals in the process of defining expectations and guidelines for a task builds commitment. Engaging the entire team in the exploratory and planning phases also creates a stronger bond between the task and individual levels, thus strengthening the team triangle.

Energy

The energy level of each team member also plays a considerable role in the strength of the individual side of the triangle. If team members are excited and motivated to tackle the activity, the outcome will naturally improve in speed and quality.

Team leaders must focus on bolstering each individual's level of energy and work to boost excitement and adrenaline whenever possible. Individual energy levels will largely be determined by how well team members feel they can do the task and how they rate their performance within the project. Focusing on each team member's strengths, celebrating victories, and praising successes can go a long way toward building team members' energy levels and improving attitudes.

Orientation toward Results

Every individual on a project team must be results oriented. Team members must focus not only on completing the task, but also on ensuring that the final result actually fulfills the purpose of the activity. If team members aren't concerned with achieving the desired results of the task, the task will be less effective and possibly pointless within the greater context of the project. In too many organizations, individual employees focus on just getting the work done because that's their job. If there is no buy-in on delivering results that achieve the purpose of the task, everyone's time is wasted.

Autonomy

Finally, individual members of the team must have the autonomy to express their creativity in how they complete the task. Team leaders must learn to delegate outcomes, not processes. If the desired results and guidelines are clearly defined and individuals are committed, energized, and focused on achieving desired results, then the team should be able to express their creativity within the process. There is always more than one way to get a job done. The individual team members ultimately own the product and own the task; therefore, they should be permitted to determine the best way to do it within the quality and requirement specifications.

In order to truly succeed as an individual unit of a high-performing team, members must have developed a critical skill: the ability to endeavor to understand, rather than require that someone else endeavor to make themselves understood. If individuals cannot learn to work toward the finish line within the defined parameters without direct oversight, they will weaken the entire team triangle.

Team

The team is the final side of this high-performing pyramid. The best teams nurture a culture of group cohesion and growth. Individuals support the team, within each task and within the project as a whole. To achieve the greatest levels of performance, teams must have a tremendous degree of openness and communication. Individual members must be on the same page for each task and the entire project, continuously identifying the team's mission and values and reinforcing ideas that have previously been developed and utilized.

The team leader must be able to help team members understand how they are going to work together on each task by providing them with examples and establishing parameters. Team leaders do not have to stop the process to focus on an issue, but rather should help the team learn how to work through issues amongst themselves without intervention. There will inevitably be conflict in a high-performance team; however, conflict is conducive to team growth and the leader should not stop the conflict. Conflict is a means of creating improved project outcomes. Every team member and team leader should become skilled in nurturing and embracing conflict for better results.

If every level of the team triangle is equally developed by a group of individuals who have been allowed to work together for long enough to form a team, the team's effectiveness will dramatically improve over time. In the preliminary development stages, teams will experience the standard forming, storming, norming, and performing phases, but eventually the individual, team, and task will meld together to form a high-performing pyramid that comes up with better, more creative ways to get the same work done and deliver higher quality results faster.

Developing the team triangle is not a short-term fix, but a long-term process that will ultimately create a more productive, efficient team—and, ultimately, an improved organization.

The Power of True Teamwork

True teamwork in action is simply astounding. Project management becomes exciting and fast-paced when a genuine team exists. Individuals who are part of a cohesive team will strive to do what they can to enhance the quality of the team, rather than focusing on their personal needs and selfish ambitions. Teammates possess laser focus on the end result, and understand that the decisions made by the team must be supported, regardless of their personal viewpoints.

When an atmosphere of teamwork is present, the members of the team appear to outsiders to be almost clairvoyant. Each member not only is able to provide other team members with the resources they need to get the job done, but also understands the issues facing each member and works to resolve those issues before they even arise. Team members can instinctively think like their teammates and anticipate their reactions to different scenarios.

Key Points

Teams can achieve true teamwork by:

Identifying and understanding each team member's strengths and weaknesses

Engaging one another in constructive conflict

Holding each team member accountable for their behaviors, decisions, and actions

Committing to group decisions, regardless of their personal viewpoints

In the face of a controversial issue, members of a true team understand what viewpoints their teammates will espouse and can enunciate these views succinctly before the debate ever begins. This insight enables the members of the team to develop solid opposing arguments before engaging in debate. A good team will also proactively work to promote only those viewpoints that will most positively impact the project—and fight passionately against perspectives that will harm the project.

Most members of teams never achieve this level of utopia, but instead quietly work together and get work done in a nonthreatening manner, avoiding conflict at all costs. To really achieve teamwork, individuals must be passionate enough about what they're doing to challenge one another's beliefs and wholeheartedly embrace the team's vision for the desired end results.

In *The Four Obsessions of an Extraordinary Executive*, Patrick Lencioni describes the characteristics of a cohesive team at great length. He says,

> Cohesive teams build trust, eliminate politics, and increase efficiency by:
>
> - Knowing one another's strengths and weaknesses
> - Openly engaging in constructive ideological conflict
> - Holding one another accountable for behaviors and actions
> - Committing to group decisions. (p. 53)

Lencioni emphasizes the importance of trust within teams, saying that "the essence of a cohesive leadership team is trust, which is marked by an absence of politics, unnecessary anxiety, and wasted energy."

Almost every project manager worth their weight will agree that eliminating anxiety, politics, and wasted energy would improve the efficiency and morale of their teams. However, many of these same managers are unable to understand the underlying cause of these problems and treat only the symptoms. This approach yields no long-lasting relief and instead ties up more of the project manager's precious time.

Key Points

When executives address the petty differences within the organization's leadership, they set an example for the team. Facing conflict head-on, in a constructive manner, demonstrates to employees that their focus should not be promoting personal ideology, but making decisions that will achieve the best results for the organization

> "Conflict is inevitable in a team ... in fact, to achieve synergistic solutions, a variety of ideas and approaches are needed. These are the ingredients for conflict."
>
> —Susan Gerke

The Cure Starts at the Top

> When things go wrong in your command, start searching for the reason in increasingly large circles around your own two feet.
>
> —General Bruce Clarke

According to Lencioni, the presence of politics, anxiety, and wasted energy is a clear indication of unresolved issues within the leadership of an organization. These unresolved issues manifest themselves in the form of office politics, and employees are left to fight their leader's battles in an attempt to gain favor as they climb the corporate ladder.

Every organization must deal with political behavior, but the organizations that succeed at eliminating this threat understand that office politics always start at the top of the food chain. Minor ideological differences between the higher-ups look like huge battles to those who follow, and because the management will rarely address those issues amongst themselves, employees are left to resolve the conflict in their own manner.

Lencioni says,

> When an executive decides not to confront a peer about a potential disagreement, he or she is dooming employees to waste time, money, and emotional energy with issues that cannot be resolved. This causes the best employees to start looking for jobs in less dysfunctional organizations, and creates an environment of disillusionment, distrust, and exhaustion for those who stay. (p. 54)

In other words, organizations whose higher-ups allow conflict to disseminate down the food chain will experience not only decreased efficiency but also excessively high turnover amongst their best employees. A project manager's job is to be the buffer in the face of office politics so that the team can get and keep the project going. The project manager must absorb the criticism in order to prevent the team from losing sight of their motivation to bring the project to fruition.

Building an environment conducive to teamwork requires that the organization's leaders quickly and effectively resolve these underlying issues. In many cases, this means that the executives must overcome their fear of conflict and engage one another in honest debate in order to establish the best direction for the firm as a whole.

Conflict can be uncomfortable, but overcoming conflict is one of the best trust-building techniques imaginable. Once leaders recognize that they can fight their battles in the boardroom and walk out as friends, they will be more willing to resolve issues before gigantic rifts tear apart the organization's leadership and spread political strife amongst their followers.

When employees see their leaders working through uncomfortable issues and overcoming personal differences for the greater good of the firm, they'll feel more comfortable and secure with the idea of engaging in constructive conflict with one another—and make no mistake, conflict is an essential component of real teamwork.

In the next chapter, we'll outline specific team-building techniques designed to facilitate true teamwork.

Chapter 3 Review

The work "teamwork" is extremely overused in modern organizations and corporations. A "team" is frequently defined as any group of people gathered to work on a specific task or activity; however, not every team is capable of teamwork. In fact, not every group of people assembled to work together on a project can be defined as a team. As the world's economy globalizes, creating a flat playing field for companies, finding the competitive edge becomes increasingly difficult. However, developing a superior team that is capable of true teamwork is one of the most powerful ways for an organization to develop a strategic differentiator over the competition.

Organizations unintentionally hinder teamwork in many ways:

- Assigning groups of people to work on an assignment together and expecting teamwork to occur magically
- Not allowing the team ample time to get to know each other and develop rapport
- Frequently reorganizing or shifting teams before they've had the opportunity to achieve true teamwork
- Allowing office politics to interfere with the highest levels of leadership within the organization

Members of a true team, also known as a "high-performing team," must understand how the other team members work and think, how team members can contribute to the project workload, and how to support and encourage the other teammates.

High-performing teams are created when every aspect of the team triangle has been developed in equal measure. If one side of the team triangle is not fully developed, the entire team will weaken and eventually fall to pieces, like a bar stool with uneven legs. The team triangle consists of three levels, each with unique needs and expectations:

1. The Task Level
   Needs:
   a. What is the expected outcome of the task?
   b. What are the guidelines for completing the task?
2. The Team Level
   Needs:
   a. Clear lines of communication between members of the team and the project leader
   b. The ability to engage in productive conflict
   c. The ability to work through problems or issues without intervention or micromanagement

3. The Individual Level

   Needs:

   a. High level of commitment
   b. Sufficient energy levels
   c. Orientation toward results
   d. Autonomy in work

4

Teams versus Groups

> Coming together is a beginning. Keeping together is progress. Working together is success.
>
> **—Henry Ford**

Clearly, not every group of people is a team—and not every team exhibits genuine teamwork.

Teamwork is a powerful tool that can be harnessed by organizations of all shapes and sizes, if the leaders of these organizations are willing to create an environment of trust, encourage healthy conflict, and build a strong team that truly works.

When all of the elements of a true team come together, the team's success—which is a reflection on the leader—can be extraordinary.

Roles and Responsibilities of the Project Manager

In *Overcoming the Five Dysfunctions of a Team: A Field Guide*, Patrick Lencioni suggests that leaders must answer four essential questions before engaging their teams in team-building exercises:

1. Are we really a team?
2. Are we ready for heavy lifting?
3. Do my team members really want to be a team?
4. Have I set the right attitude and laid the groundwork for safely becoming a team?

Are We Really a Team?

In order to answer this question, the leader must determine whether his team is really a team or just a group. Lencioni defines a "team" as a small group of people that "shares common goals as well as the rewards and responsibilities for achieving them. Team members readily set aside their individual or personal needs for the greater good of the group."

As you saw in chapter 3, team members must be willing to subjugate their personal beliefs in favor of the viewpoints that are most favorable to the group and the desired results of the project. If team members are unwilling to do this, conflict shifts from healthy debate to unnecessary infighting—and the environment of trust is lost.

If you determine that your team is really just a group, look within the group to find a smaller subset of people that can actually achieve true "team" status and work with that group of individuals.

Are We Ready for Heavy Lifting?

Once you've determined that your group is capable of true teamwork, you must decide whether the members are prepared to put forth the energy and effort required to form a successful team.

Team-building exercises require members to step outside of their comfort zone. Members must be willing to drop some of their social filters to familiarize themselves with their teammates on professional and personal levels. There is no shortcut for familiarity. Individuals must spend time working with their teammates in order to understand how they think and how they operate.

Please note that developing deep familiarity between members of a team is a process that needs to be eased into. I once worked with a group who tried getting personal too quickly, and a prominent member of the team admitted something about his marriage that made everyone else feel extremely uncomfortable. Rather than helping everyone open up and share personal ideas, this team member's admission shut everyone else down because the sensitive personal information was offered up too early in the process. When completing trust-building exercises, the team must begin with a series of relatively safe questions, before progressing into less and less safe questions. This helps the team develop a small degree of trust before tackling more difficult questions.

Becoming a team does not have to take years or even months, but understand that your group will not make the transition overnight. The journey toward true teamwork is not short or effortless, but the rewards at the end of the path are numerous. To succeed, members must be committed to the process and focused on the results.

Assessing your team prior to embarking on a team-building exercise is a smart move. Assessment should be done in three stages:

1. *Conduct an assessment of the needs and capabilities of each individual on the team.* This research is easier said than done. Bringing in a consultant to assess each individual adds third-party bias into the equation, and reviewing documentation may not provide a complete picture of each member's capabilities. The only way to really understand the abilities of team members is to meet with each individual one-on-one to determine the following:

- What value the team member believes he or she brings to the team (i.e., what role does the team member feel best suited for?)
- Strengths and weaknesses
- Leadership needs (what does the team member need from you?)
- Personal interests
- Primary motivations (money, family, personal fulfillment, etc.)

2. *Once you understand an individual's interests, motivations, and roles from his or her perspective, provide each team member with a specific task that relates to his or her role within the project.* Make certain that each member understands the agreed-upon result of the task. The tasks do not have to be complex; you just need to assess whether they can achieve the required results in the designated time frame. This assignment is an excellent preliminary step in understanding your team and developing a baseline to monitor their progress throughout the team-building exercises.

3. *Evaluate the current strength of your team.* The following questions can be used to guide your assessment:
   - Are the members of my team willing to engage one another in genuine confrontation and debate?
   - Do team members step outside of their comfort zone to defend their beliefs about the best course of action?
   - Are meetings used for show-and-tell or to address essential issues?
   - Do members of my team understand and respect one another?

For some strategies on how to use assessment within your team, please go to our companion website, www.ManagementToLeadership.com.

Do My Team Members Really Want to Be a Team?

Have you ever seen management put together a "dream team" for some project, only to watch as that team failed miserably? Why does that happen?

Having the right people in place and telling them to work together are necessary steps in forming a team. Gathering your people together forms a group—but putting together a group of stellar individuals is not sufficient for making a team.

Why do dream teams sometimes fail? There are a number of reasons. Sometimes they fail because of personality conflicts. Sometimes the stars are all vying with each other to be the superstar or leader. Sometimes team members have different styles or methods that are just incompatible (and they are loath to learn new styles or methods). And sometimes the problem is with the team leader alone.

The crucial metric for a team, then, is not the abilities of the individuals, but the dynamics of the relationships between them. Your team must be willing to work together. They must want to be a team.

Deciding whether or not your team members want to be a team is tricky— if you ask them outright, they are likely to pay lip service to the idea of a team. So, you will have to look for symptoms of poor team cohesion as you watch the ways in which your team members interact:

- Do team members chat at all during unstructured time (before meetings, during coffee breaks, after lunch, etc.)? Do they ever discuss ideas, problems, or projects that they have? Or does everything seem overly formal?
- Do team members seem willing to work on each other's challenges and problems? Or do they use others' problems to their own advantage?
- Do team members take responsibility for results? Do they take responsibility for mistakes? Or are they in a perpetual "pass the buck" mode?
- Do team members keep track of their own successes? The successes of others? Or are they just going though the motions most of the time?

If your answers are more in line with the descriptions after each "or," these are some sure signs that your team members aren't interested in being a team at all.

Have I Set the Right Attitude and Laid the Groundwork for Safely Becoming a Team?

This question is not so much about your team as about yourself. You will need to set the right tone and attitude from the very beginning—perhaps even before the team is formed. When you do form your team (or become team leader, if the team is already formed), you will need to set some procedures and ground rules for interaction. Two exercises in the appendix, "Clarification of Organizational Principles" and "Clarification of Team Principles," would be good to do with your team in order to set the right attitude and lay the groundwork for becoming a successful team.

Encouraging True Teamwork

One of the best ways to determine whether or not true teamwork is present within a group is to observe a meeting. In many organizations, meetings are dreaded tasks: boring sessions in which little is accomplished or nothing of value is said.

That's not the case when a true team sits down for a meeting. Lencioni says,

> For cohesive teams, meetings are compelling and vital. They are forums for asking difficult questions, challenging one another's ideas, and ultimately arriving at decisions that everyone agrees to support and adhere to, in the best interests of the company. (61)

If team members are checking their e-mail, text messaging, daydreaming, or otherwise disconnecting from the task at hand, the team has a major problem. True teamwork requires that every member be involved and alert, ready to hold their teammates accountable when the need arises. Every member must be geared up for a fight.

Redefining Conflict

> Good leadership requires you to surround yourself with people of diverse perspectives who can disagree with you without fear of retaliation.
>
> **—Doris Kearns Goodwin**

The word "fight" may conjure images of red-faced men in suits angrily pointing fingers at one another as they insult each other's ideas, character, and behaviors from across the table. That's not the type of fighting any team needs or wants; in fact, this type of fighting is detrimental to the team's health and productivity and should be discouraged at all costs. The type of conflict that exists in high-performing teams is a valuable tool to spur innovation, efficiency, and quality decision making. This conflict, which can be defined as a productive conflict, differs from traditional conflict in many ways.

Productive Conflict Is about Getting to the Heart of the Issues, Not Getting in a Jab at Someone Else

Teammates do fight—but not on a personal level. Teams fight heatedly about the issues, and wholeheartedly espouse the beliefs that they believe are in the best interests of the team and the project. Members use the debate as an exploratory process, and challenge one another to defend their beliefs and assumptions.

People Promote the Idea That Is Best for Their Team, Not Their Favorite or Personal Opinion

Conflict becomes a fact-finding mission, and once both sides have presented their case, the group makes a decision based on the best interests of the team and the project. The viewpoint that each individual promotes in the end is not necessarily congruent with that individual's personal viewpoint about the project. Instead, each member will promote the views and ideas that they believe will best achieve the goals of the team.

Team Members Can Argue Both Sides to Aid in the Exploratory Process

It's not uncommon to see team members switching sides repeatedly throughout the course of the debate. This flip-flopping is not a bad thing, but rather is a testament to the merits of both solutions being debated.

Everyone Accepts and Stands behind the Final Decision of the Team, No Matter What Their Personal Preference

Once the team has made a final decision, every member of the team embraces the decision, regardless of which position he or she took in the debate. Most importantly, there are no hard feelings when the members exit the boardroom.

Achieving a level of comfort and familiarity that allows for genuine conflict is not an easy task. One way to help your team become comfortable with the idea of productive conflict is to establish a team charter agreement at the start of the project to determine how to deal with conflict. Working together to establish the ground rules keeps everyone on the same page, and establishes expectations for healthy conflict between team members. Once the charter has been agreed upon and recorded, the charter should be displayed prominently in the work area and referred to whenever necessary.

Hitting below the Belt

An important note about conflict: in true teams, members are cautious to keep their comments directly related to the issue at hand and avoid crossing the line into unrelated personal issues. Yes, there is the possibility that someone will get caught up in the heat of the moment and make a personal attack, but the difference in these situations is that an atmosphere of trust has been cultivated.

When a member of a high-performing team has a lapse of judgment and hits below the belt, the other members of the team will work together to move past the encounter and set things back on track. The team will work together to address the unfair remarks, and no one will leave the room until all hurt feelings have been addressed and the entire room is prepared to forgive and forget.

It's important to note that even productive conflict should never be encouraged just for the sake of arguing. True teams do not engage in passionate debates over petty, insignificant issues. Passionate debate is reserved for only those issues that require the input of the entire team and a thorough examination of the issues. Not every issue is worth fighting for; true teams know how to pick their battles wisely.

A good team is a group that has developed a bond of trust and mutual respect. Members of a true team will recognize that other individuals may feel uncomfortable or downright hostile to certain ideas, but will work to integrate these individuals into the decision-making and implementing process rather than alienate dissenters as most groups do. This is not easy work, but effective project managers understand that genuine conflict is essential.

One wise project manager I knew once said to a group of eight,

> If we all agree about every issue, we have no need for seven of our members and I'll get rid of all of you. Dissenting viewpoints are not encouraged—they are required. I don't need any more "yes-men." For heaven's sake, don't be afraid to tell me what you really think, even if your opinion is going to really tick me off.

Then the manager smiled, and his team stepped up to the challenge and began questioning the previous consensus—because their leader had created an environment of trust. Team members knew that they were not going to be penalized or ridiculed for expressing their opinions and felt comfortable sharing their beliefs.

Developing Your Team's IDEAS

IDEAS is an acronym used to describe the requirements of true teamwork. If teams want to grow together, create an environment of trust, and improve their efficiency, they must apply themselves to enforcing IDEAS:

Identifying attributes

Debating essential issues

Embracing accountability

Achieving commitment

Setting and maintaining high standards

Identifying Attributes

For the purpose of this section, the word "attributes" is used as a catchall term to describe the unique variety of traits, habits, attitudes, and behaviors that make each individual unique.

Specifically, team members must understand one another's "whats," "whys," and "hows."

- *What the person does or is*: determined by identifying each team member's major strengths and weaknesses
- *Why the person is the way that he or she is*: determined by understanding each team member's personal history
- *How these inherent characteristics manifest themselves as actions*: determined by examining the results of each team member's behavioral profile

Identifying attributes aids in building that essential environment of trust within a team. When team members can identify "what" attributes their teammates possess, "why" they are this way, and "how" these attributes manifest themselves, the team will become more comfortable with one another.

Members will be more willing to share their true beliefs and ideas without fear of ridicule or ostracism because they've already "gotten naked." Because their inner attributes have already been exposed, team members will have nothing left to hide, thus eliminating unnecessary pretensions and posturing.

Determining the what, why, and how of team members' attributes can be achieved by completing the three simple exercises in the "Identifying Attributes" section of the appendix:

1. The 360 Review
2. Personal Histories
3. Behavioral Profiles

These exercises are powerful tools in helping teams to overcome the fundamental attribution error. Lencioni defines this error as follows:

> Human beings tend to falsely attribute the negative behaviors of others to their character (an internal attribution), while they attribute their own negative behaviors to their environment (an external attribution). (64)

People generally understand their own attributes. This insight allows individuals to "cut themselves some slack" in the face of a poor judgment call or hurtful action. Unfortunately, we often do not understand the intricacies of our coworkers' behaviors and mind-sets well enough to extend them the same courtesy.

Overcoming the Fundamental Attribution Error

Most people are able to shift the blame to their environment, rather than their character when they mess up.

However, people tend to assume the worst about others in the face of questionable behavior. The offending behavior becomes the rule, rather than the exception, and the individual's character is deemed to be flawed.

An understanding of the "what, why, and how," of each team member's attributes can help foster an environment of empathy rather than judgment.

When team members do not have a clear picture of each other's attributes, they tend to automatically assume the worst in the face of questionable behavior. They aren't willing to believe that the action was an exception rather than the rule, and make harsh (and often lasting) assumptions about the offender.

The "Identifying Attributes" exercises help team members to understand one another on a deeper level, reducing the likelihood that teammates will jump to unfair conclusions about each other in the face of conflicting views or actions.

Although the exercises might appear to be "soft" or "touchy-feely," these are powerful tools for helping teams foster an environment of trust and empathy, which is essential to the next step in IDEAS: fanning the flames of conflict and debating essential issues.

Debating Essential Issues

The importance of conflict—passionate debate about essential issues surrounding a project—was covered in detail in Chapter 3.

To recap, examine the five criteria of healthy conflict:

1. Conflict is used as a fact-finding tool. Each team member presents his or her opinion of the best solution and works to identify gaps or inconsistencies in other proposed solutions.

2. The promotion of a particular viewpoint should not be made based on personal reasons. Teammates should espouse the perspective that they feel is most congruent with the desired outcome of the project.

3. Teams do not allow any hitting below the belt; personal attacks are strictly forbidden. All claims should pertain specifically to the issue—not the characteristics of the person making the argument.

4. No fighting for the sake of fighting; conflict is limited to important issues that directly affect the project and its desired outcome.

5. Once the team has committed to a course of action, every member of the team must commit to that decision, regardless of their personal beliefs about the issue.

Specific exercises to aid in encouraging constructive conflict can be found in the "Debating Essential Issues" section of the Appendix:

1. Conflict Resolution Exercise
2. Conflict Continuum Exercise
3. Thomas–Kilmann Model

Embracing Accountability

Humans tend to paint themselves in the best light possible while throwing any hint of doubt or culpability toward others. But the most successful teams build strong bonds that prevent such "buck passing" from happening.

The exercise called "Team Scoreboard" is a tool that you can use to track a team's progress and keep them accountable by posting results. In addition to this exercise, you might try the "Team Effectiveness Exercise" in order to generate both positive and critical feedback.

The most important thing to keep in mind for embracing accountability is to keep everything explicit and open. For example, every team member should know the following:

- What the team is about: the aims, vision, and goals of the team
- What he or she is personally responsible for
- The metrics used to measure team success
- The metrics used to measure individual success
- The correct person to see for common problems
- The standard grievances process
- The actions that require go-ahead from the team leader

Write out these policies ahead of time, and refer to them often.

Achieving Commitment

Team members need to be committed in order to achieve the best results. They should be able to defer to team decisions and be willing to take on the team's projects and challenges as their own.

Commitment is not consensus. Consensus is incredibly difficult to achieve for any group—and nearly impossible for a group where talents, styles, and opinions differ. Waiting for consensus on any issue will likely stall a team before anything worthwhile can be achieved.

Commitment is the ability to stick by a decision despite a natural tendency to disagree. Recall Al Gore's words after losing the 2000 presidential election: He said of George W. Bush "He's my president too," meaning that, although they disagreed about every issue (including who should have won the election!), Gore was committed to the process and willingly recognized Bush as the new head of state.

Commitment is not something that can be created or taught. Commitment must be "grown" in each team member over time.

So how does one ensure that the members of a team have commitment, if this is a quality that cannot be created or taught? The best way to achieve a high level of commitment in your team is to make your team members a part of the process before work even begins. Give them input on everything from team vision to e-mail policies. By involving your team in the development of every process, you give individuals a sense of owning the team and make feelings of commitment much more likely.

As mentioned, the Appendix outlines two exercises for involving team members in policy decisions: "Clarification of Organizational Principles" and "Clarification of Team Principles." If possible, try doing these exercises as soon as the team is formed.

Setting and Maintaining High Standards

Roger von Oeck, creativity guru, best outlined the importance of high standards: "Hit the bull's eye every time? Well then maybe you're standing too close to the target."

By setting high standards for your team, you push team members farther than they might normally go. They might not hit the target—but they will get a lot closer than you expected. When you set the bar high for your team, you encourage them to stretch just beyond what they think they can do. Many professional coaches say that the best goals require a little bit of reach—the objective should be defined as just beyond what you know the team is capable of.

Remember, you want your target—your standards and objectives—to be concrete and achievable, not abstract or unrealistic. One good way to set standards is with the "Resumé Exercise" found in the Appendix. In this exercise, you have your team compose an ideal resumé of a fictional employee, and then use the resumé as a template for excellence in your own team.

Another way to encourage your team to aim for difficult goals is to give them team-building exercises like "The Tower of Babel," also in the Appendix. Such exercises take some time and space to run, however, and so are probably best saved for an off-site workshop.

The IDEAS are key steps in establishing trust and commitment, which naturally lead to efficiency and personal growth. Notice that the role of the team leader in each of these steps is not to dictate policy or goad employees into working a particular way. At best, the team leader simply acts as a kind of facilitator that lets the team develop in a natural way. Remember, these are the team's IDEAS, not your own!

Chapter 4 Review

Not every team is capable of teamwork; some "teams" are merely groups assembled to work together on a task or activity. Whether or not a group evolves into a high-performing team depends largely upon the abilities of the project leader. Before embarking on a project with a team, a project leader must determine several things:

1. Is the assembled group really a team?
2. Is the team ready for hard work?
3. Does the team want to be a team?
4. Has the proper groundwork for becoming a team been developed?

Not all groups are capable of teamwork, nor do all groups want to put the time, effort, energy, and emotions into developing true teamwork. Assessing the team as a whole and each team member individually is critical to the development of a high-performing team.

Encouraging Productive Conflict

A team's approach to conflict is one of the biggest differentiators between ordinary groups and high-performing teams. First and foremost, the presence of conflict is essential. True teams must be willing to debate and argue the issues in an appropriate manner. True teams engage in productive conflict to gain enlightenment on the issues and determine the best course of action—they don't just fight for the sake of fighting.

Productive conflict differs from fighting in a number of ways:

- Productive conflict is intended to get to the heart of the matter.
- Productive conflict is not about getting in a jab at another team member.
- Team members argue for the idea they believe is best for the team, not their personal preference.

- In productive conflict, members of the team can argue both sides of the coin.
- At the conclusion of the debate, members of a true team unite behind the team's decision, regardless of which side they promoted during the conflict.
- Hitting "below the belt" is discouraged, and any personal attacks or comments are directly addressed.

Teamwork IDEAS

The characteristics of a true team can be summarized through the acronym IDEAS:

- Identifying attributes (What are the team members' strengths and weaknesses?)
- Debating essential issues (Can the team engage in productive conflict?)
- Embracing accountability (Do members of the team take responsibility for their actions?)
- Achieving commitment (Are members of the team dedicated to and supportive of the team's goals, mission, and values?)
- Setting and maintaining standards (Has the team leader established a baseline for success? Do members of the team accept and reach to achieve these standards?)

5

Leadership versus Management

Effective leadership is putting first things first. Effective management is discipline, carrying it out.

—**Stephen Covey**

Many corporate executives use the terms "leadership" and "management" interchangeably, as if one were synonymous with the other. This is a mistake.

The difference between a leader and a manager can be easily spotted in the hiring process. You can hire an excellent manager in a week, but you might spend years looking for a leader of the same caliber.

Management is about the tasks: planning, budgeting, organizing, and problem solving. A manager's role is to ensure that the business's vital functions are completed and that numbers are maintained. A manager's role is essential to a company's survival, but not more so than the role of a leader.

Leadership moves beyond the tactical elements of organization, individuals, and getting things done. If management is about the tactics, then leadership is about innovation. Leadership is about seeing where the team needs to go and guiding the project in that direction.

Leadership is also about people. Organizational psychologists believe that most people have the ability to develop relationships and nurture friendships with anywhere between 100 and 150 people, geographic proximity permitting. Regretfully, many project managers are assigned projects that involve many more than 150 people, and those people may be scattered across the globe in multiple time zones.

In these cases, the leader must be able to step out of a direct management role and promote leadership, vision, and understanding throughout the team to get the project done. Therefore, leadership is about the character and abilities of the person who leads.

Key Points

- The role of a leader and the role of a manager are not one and the same. Leadership refers to the leader's ability to influence, guide, and inspire his or her team in order to get the job done.
- Leadership is not about personality or style; leadership is about substance and the leader's character.

Leadership is character in action.

—**Warren Bennis**

A leader's job is to motivate others to get the job done, and get the job done right. A leader's job is to inspire his or her people to take action, and shape employees into better workers (and future leaders). A leader's job is to provide his or her people with the tools necessary to tackle the items on the manager's list.

There are many great managers in this world. Many of them are superb at one thing: reading a balance sheet, creating strategies and visions, coordinating financial deals, or forecasting earnings. While these are admirable skills, many of those same managers are unable to influence their teams to do much of anything. These managers couldn't even lead a horse to water—much less make the animal take a drink.

One of the reasons that the word "management" is still used is because leadership is difficult to define and evaluate. Developing leadership takes time and energy, and the merits of a leader are difficult to quantify, particularly in the short term. Management techniques, on the other hand, are designed to be easy to quantify and direct, and therefore easy to manage (by the higher-ups).

Leadership is not about planning, budgeting, organizing, and forecasting. Leadership is about the attitude and character with which you do these things, or the way you influence others to do those things.

While the terms "leadership" and "management" are not one and the same, good leaders do tend to be good managers because they understand how to get the job done—if not by themselves, then through their followers.

So what is the difference between a leader and a manager? John Maxwell explained the distinction best when he said, "Leadership is influence. Nothing more, nothing less."

A manager is given a series of tasks that he or she must accomplish by working with other people. A leader is given (or must create) a vision, and must achieve that vision by working through other people and influencing them to come on board with the team.

The leader recognizes that his or her followers are not just cogs in the machine, that each individual will have either an interest or disinterest in supporting the vision. If the individual is not inclined to support the mission, the leader's role becomes to inspire and motivate that person to embrace the vision.

Leadership is not about what you do. Leadership is about who you are. Leadership is about what you inspire others to be.

Leadership Styles

Although leadership is about the person, leadership is not about personality. There is not one defining personality trait that separates leaders from followers; therefore, leadership is not something you're born with. Leadership is a trait that you can methodically develop and nurture until leading comes naturally to you. That's good news.

History's greatest leaders fall across a vast spectrum of varying personalities. For example, consider the differences between each person in the following pairs:

- Generals Dwight D. Eisenhower and George Patton
- Presidents Ronald Reagan and John F. Kennedy
- Martin Luther King Jr. and Malcolm X
- Mother Teresa and the Dalai Lama

Each of these leaders was (or, in the case of the Dalai Lama, is) famed for his or her ability to lead. While these leaders possessed remarkably different leadership "styles," the differences in the leader's personalities are insignificant. We remember these leaders because of their incredible ability to influence and inspire their followers to achieve monumental accomplishments.

Leaders may be loud, shy, short, thin, fat, old, young, unkempt, stylish, articulate, awkward, boring, or exhilarating. You might say a leader can be anything at all. So, rather than examining the differences between different leaders' styles and personalities, one would do better to examine the common thread found in all great leaders.

What characteristics do all great leaders share?

- Great leaders do not spend their time determining, "What's in it for me?" They aren't squabbling for the corner office or engaging in office politics in an attempt to get ahead. Great leaders are not micromanaging. In fact, great leaders don't even spend much time thinking about the tasks associated with management: numbers, facts, and figures.
- Great leaders don't spend sleepless nights worrying about whether or not their personal needs are being met—they stay up worrying about the needs of their teams.
- Great leaders are humble.
- Great leaders serve their people.

The Servant Leader

The servant-leader is servant first. It begins with the natural feeling that one wants to serve, to serve first. Then conscious choice brings one to aspire to lead. That person is sharply different from one who is leader first, perhaps because of the need to assuage an unusual power drive or to acquire material possessions....The leader-first and the servant-first are two extreme types. Between them there are shadings and blends that are part of the infinite variety of human nature.

—Robert K. Greenleaf (73)

The outdated, dictatorial modes of leadership are giving way to a better form of leadership, a model that emphasizes the importance of teamwork, ethical behavior, team decision making, and personal growth. Developed by Robert K. Greenleaf in 1970, servant leadership is one of the most popular and effective leadership models in existence today. Thirty-five of *Fortune* magazine's "100 Best Companies to Work For" practice the principles of servant leadership. The premise of servant leadership is this: the servant leader serves the people he or she leads, viewing them as an end in themselves rather than a means to a corporate purpose or bigger bottom line.

Servant leaders understand that they must meet the needs of their followers in order to enable them to meet their full potential. Servant leaders are not self-serving, domineering leaders who use their power to force their followers into following commands. Rather, servant leaders strive to respect and motivate their followers, using their influence to inspire followers to reach new heights.

In *The Road Less Traveled*, M. Scott Peck said,

> Servant leadership is more than a concept. As far as I am concerned, it is a fact. I would simply define it by saying that any great leader, by which I also mean an ethical leader of any group, will see herself or himself as a servant of that group and will act accordingly. (p. 73)

Leading from Behind

We often depict leaders as being the face and the voice of the team, the person out in front leading the rest along. When visualizing the act of leading, one would likely imagine the leader as being in the front of the group, literally guiding the team as if on a journey. Leaders are often thought of as being in the front of the room delivering speeches or at the head of the table running a meeting. The leader is the front person, the chief, the person in charge, and the boss.

Servant leaders, however, do not guide their team by being in front and literally leading the way. They lead their team from behind. Rather than plunging forward as the rest of the team follows, servant leaders lead the project by serving with and amongst their team, offering help and support so that the project can move forward.

If you compare the completion of a project with a race, the servant leader would not be the first to the finish line, ready to accept the trophy and collect the praise for the team's success. Instead, the servant leader would cross the line in lockstep with the last member of the team, offering words of support and encouragement through the last steps. The servant leader understands that his or her success as a leader is contingent on the success of the team, and that the team is only as strong as its individual members. Servant leaders understand their team members' unique strengths and weaknesses and consider those characteristics when making decisions with and for the team.

If you are leading your team from behind, that doesn't mean that there is no one up in front forging the team ahead. Part of leading from behind requires that you allow your followers to take their turn at the helm serving as the team leader. Servant leadership involves serving your team and giving them the tools to take the lead.

In an article in the *Harvard Business Review* titled "Where Will We Find Tomorrow's Leaders?" Linda A. Hill, a professor of business administration at Harvard Business School, says that leading from behind allows a team to exercise its creativity in new ways by stimulating innovation and discovery. "You have to create an environment in which (members of your team are) engaged and in which the collective talent of team members is tapped by having everyone take the lead at some point."

Hill drew a poignant example of leading from behind from renowned servant leader Nelson Mandela's autobiography, *Walk to Freedom*. In the book, Mandela describes how leadership was defined to him by a leader of his tribe. "A leader, he said, is like a shepherd. He stays behind the flock, letting the most nimble go out ahead, whereupon the others follow, not realizing that all along they are being directed from behind."

As the old saying goes, if you are leading from behind, you cannot block your team's collective vision. In an interview with the *Dallas Business Journal* Dennis Bounds, president of the Bill J. Priest Institute for Economic Development in Dallas, says that companies must rethink their approach to leadership if they want to thrive and rise above the competition. In order for an organization to create high-performing teams, improve workflow systems, attract high-caliber recruits, and improve the bottom line, organizations must move away from "the old fashioned approach of barking orders and expecting employees to carry them out."

"Leadership is a lot more than just leading an organization," Bounds said. "It's inspiring and engaging an entire organization so that it becomes a 'leader-full' organization, where everyone is a leader" (p. 74).

Leading from behind while your team serves out in front causes team members to take ownership of their role and will encourage employees to innovate and brainstorm ways to improve their performance. If you let the members of your team take control in unique situations based on their strengths and talents, you will learn more about your team and experience better results. You'll also have more time to focus on specific problems within the team and project—more time to serve the needs of your team to ensure that they're performing at their highest level of performance.

Characteristics of a Servant Leader

In *The Servant Leader Within,* Larry C. Spears, CEO of the Robert K. Greenleaf Center for Servant-Leadership, outlines the ten characteristics or skills of a true servant leader:

1. *A servant leader is a good listener.* Servant leaders are committed to listening to others in order to both identify and clarify the will of a group. They pay attention to what their followers are saying, and intuitively understand their unspoken needs. By truly listening and reflecting on their followers' thoughts and desires, servant leaders are better able to meet the needs of their team.

2. *Servant leaders are empathetic.* Servant leaders work to empathize with their followers. By adopting an attitude of benevolent distortion, servant leaders are able to see the good intentions of their followers, even when their performance is unacceptable. This understanding enables the servant leader to correct problematic behaviors without persecuting the person—to address the incorrect actions without rejecting the follower.

3. *A servant leader is a healer.* According to Spears, "Learning to heal is a powerful force for transformation and integration." Human beings are, by their very nature, flawed and often carry emotional wounds with them into the workplace. A servant leader works to make his or her followers whole again by helping them to grow and evolve as individuals.

4. *Servant leaders are aware.* By being aware of their surroundings and the needs, wants, and desires of their followers, servant leaders are able to understand the complex issues surrounding their followers' ethics and values and view these issues from a position of neutrality. The servant leader also possesses a deep awareness of self, and constantly works to address his or her own shortcomings and misconceptions.

5. *The servant leader is persuasive.* Servant leaders use influence, and not authority, to lead their followers. Rather than forcing followers to comply with their demands, servant leaders strive to persuade their colleagues to follow their lead.

6. *Servant leaders are able to conceptualize.* Servant leaders are able to look beyond the day-to-day realities of life and see the big picture. While maintaining a focus on short-term goals and objectives, the servant leader also expands his or her thinking to include a conceptual approach to the organization's long-term goals and mission.

7. *Servant leaders possess great foresight.* A servant leader has the uncanny ability to predict the likely outcome of a situation. By weighing lessons learned with present realities and the potential consequences of each decision, the servant leader can foresee the likely chain of events that stems from specific circumstances.

8. *The servant leader is a steward.* Servant leaders' primary commitment is to serve the needs of others in order to help them reach their full potential and the organization's goals. Servant leaders are stewards for the greater good of society.

9. *Servant leaders are committed to the growth of their people.* Servant leaders do not see people as a means to an end, but rather as the end in and of itself. A servant leader strives to nurture the personal, professional, and spiritual growth of his or her team. Spears wrote that "servant leaders believe that people have an intrinsic value beyond their tangible contributions as workers. As such, the servant leader is deeply committed to the growth of each and every individual within his or her institution" (p. 76).

10. *The servant leader works to build community.* The servant leadership model concludes that a sense of community can be established among the employees of an organization. Servant leaders work to find ways to create a sense of community for an organization's workers, filling a void left in the shift from local communities to corporations as the primary influence in employees' lives.

Servant leaders serve their followers before they attempt to lead. Servant leaders strive to extend and improve themselves in order to lead their followers as best they can. Great leaders work to raise their game, improve their character, and become better people so that they can better serve those around them.

A servant leader might encourage personal development and a sense of community within his or her team by doing the following:

- Establishing a fund for the team's personal and professional development
- Demonstrating interest in every team member's ideas and suggestions
- Actively involving team members in the decision-making process
- Helping laid-off workers find new employment
- Addressing a team member's personal struggles rather than insisting that they "check their personal issues at the door"

Although servant leaders' first and foremost objective is to serve their followers' needs, servant leaders are not slaves to their people.

Wants versus Needs

James Hunter makes an important distinction between servant leadership and slave leadership. Slave leadership involves meeting people's wants, while servant leadership involves meeting people's needs.

want

> v.
>
> 1. To have a strong desire for; to have an inclination to
> 2. To wish or demand the presence of

need

> n.
>
> 1. A legitimate physical or psychological requirement for the health and welfare of human beings

The difference between needs and wants is monumental. You will never be able to give your people everything that they want, nor would I advocate that you try. However, you do have a responsibility to provide your people with the things they need, particularly in regard to what they need to get the job done.

The difference between needs and wants applies to how you approach your own personal needs and desires as well. Servant leaders must be willing to subjugate their own personal needs, wants, moods, and issues in order to better serve their followers' needs.

Leader may not want to discipline their followers; a given leader may be a naturally kind and good-hearted person who struggles with the idea of reprimanding another human being. Nonetheless, that leader must be able to overcome his or her personal inclinations (wants) and give team members what they need, including administering appropriate discipline where necessary.

Subjugating your personal desires to meet the needs of your team requires selflessness and a willingness to extend yourself beyond your comfort zone to serve your followers.

Power versus Influence

> Being powerful is like being a lady. If you have to tell people you are, you aren't.

> **—Margaret Thatcher**

A project manager who does not understand the distinction between power and influence will never be able to lead. For our purposes here, you can assume leadership is synonymous with influence—to lead is to influence.

---

pow-er

n.

1. Strength or force exerted or capable of being exerted; might
2. The ability or official capacity to exercise control

in-flu-ence

n.

1. The act or power of producing an effect without apparent exertion of force or direct exercise of command
2. The power or capacity of causing an effect in direct or intangible ways

---

Power, on the other hand, is "Do what I say or you're fired." Power is forcing someone to do something simply because you rank higher on the food chain. Power might get the job done for a while, but eventually power will erode your relationships with your people.

Remember: power can be bought, sold, given—and taken away. People can obtain power through family ties, politics, or imitation.

One need look no further than a history book to find examples of "leaders" who ruled with iron fists of power. Many of these monarchs, czars, and tyrants were eventually overthrown by civilians with enough influence to persuade others to say, "We've had enough."

Max Weber once said, "Authority is the skill of getting people to willingly do your will, because of your personal influence." In other words, people do what you want or need them to do not because of your power, but because of your influence.

Consider the power of a mother whose adult son has moved out of the family home. If the mother makes a request of her son, she doesn't have much power to ensure that he does what she asked. After all, if she sends him to his room, the son can just return to his own home. The threat of withholding an allowance is similarly lost. The mother's power over her son is depleted, yet her influence remains. Incredibly, influence can actually create more power than power itself.

When a leader has to exercise power to get things done, his influence, and thus his leadership, have faltered.

In *Becoming a Person of Influence*, John Maxwell and Jim Dornan highlight the attributes of a great influencer using the word "influence" as an acronym.

Integrity	Builds relationships on trust
Nurturing	Cares about people as individuals
Faith	Believes in people
Listening	Values what others have to say
Understanding	Sees from their point of view
Enlarging	Helps others become bigger
Navigating	Assists others through difficulties
Connecting	Initiates positive relationships
Empowering	Gives others the power to lead

Influence and Authority

Influence, like leadership, is about who you are. Influence cannot be bought, sold, traded, or taken away. The only person capable of limiting your influence as a leader is yourself.

Gandhi is a perfect example of the strength found in influence. Gandhi certainly had no power, but because of his influence, he convinced the British Empire to give up India, their most profitable providence, without ever firing a single weapon.

Gandhi developed authority through service and sacrifice. Gandhi was dedicated to serving his people and his cause. He was beaten and imprisoned. He fasted. In the end, Gandhi made the ultimate sacrifice: his life.

Gandhi also got people's attention. Gandhi's followers paid attention when he spoke. He never resorted to violence or scare tactics, but obtained worldwide recognition by sacrificing himself for his cause. Gandhi was given the largest state funeral for an individual who held no official office in India's history. Gandhi had such influence over and respect from his followers around the world that to this day, his house is kept as a shrine—not to religion, but to the great man himself.

In the end, Gandhi's authority—gained through service and sacrifice—defeated the power of the great British Empire.

Leadership is built upon influence, and influence is best obtained through service and sacrifice. Influence cannot be conferred upon you; you must become an influencer.

The Ultimate Test of Servant Leadership

> The best test, and the one most difficult to administer, is: Do those served grow as persons? Do they, while being served, become healthier, wiser, freer, more autonomous, more likely themselves to become servants? And, what is the effect on the least privileged in society? Will they benefit or at least not be further deprived?
>
> **—Robert K. Greenleaf (80)**

As Greenleaf states in the oft-repeated quote above, the ultimate test of leadership is whether your team grows under your tenure. Your success as a servant leader isn't determined by the quality or caliber of your team now, but by the depth of their improvement when they leave your guidance. Servant leaders serve their teams so that they can grow as a group and as individuals. This requires the type of coaching and discipline that can only be provided by a leader who is leading from behind and working in the trenches with their teams.

Many organizations cling to the notion of yearly performance reviews, where the management sits employees down and tells them what the organization thinks about how they're doing. Performance reviews shouldn't be a formal, annual event, but instead a regular occurrence in which the leader offers team members specific guidance and direction as to how they can better meet and exceed the organization's expectations. If a servant leader is truly working with and through his or her team, these performance reviews will occur at spontaneous times, when the lesson is truly needed. When the servant leader witnesses a positive or negative act, the appropriate response should be delivered in that immediate moment for greatest effect.

As with training an animal, the timing of a servant leader's praise or corrective actions is essential. In the sports world, coaches don't just hang out on the bench for the entire season, taking notes about each player's performance. The coach is out there for every play of the game, adjusting team members' positions, giving them support and encouragement, and addressing less-than-stellar performances on the spot. The coach is in the game with his or her team, helping them be the best players they can be. As a servant leader, you can empower your team and work through them to reach new heights by leading at a new level, one that moves beyond active to a truly integrated level of leadership.

Have you ever experienced what it is like to be led by a servant leader? It is likely that you have, even though you may not have recognized the servant leader as a leader at the time. In many cases, the servant leader may not have been the person you would have thought of as "the leader." Many servant leaders are serving just out of the limelight, away from the praise and the center of attention. Yet these behind-the-scenes people are the ones pulling the puppet strings to make things happen.

Sometimes, the servant leader is in the role of an administrative assistant or assistant manager. While the "leader" is off with the higher-ups advancing his or her personal career, the servant leader emerges to ensure the team meets its goals. Other times, the servant leader is in the form of a mother, teacher, coach, or caretaker who motivates his or her followers to be the best they can be.

Servant leaders have a subtle way of making sure that what needs to be done gets done, without ever seeming to give orders. These intuitive leaders have a sense of what each person needs to contribute, and understand the best way to communicate with each unique member of the team. Servant leaders lead by serving their team, which in turn ensures that the team is able to reach a successful conclusion.

Does Your Organization Seek Enduring Success or Eventual Extinction?

In *Insights on Leadership*, Stephen Covey emphasizes that the concept of servant leadership is one that is here to stay.

> I am convinced that (servant leadership) will continue to dramatically increase in its relevance. You've got to produce more for less, and with greater speed than you've ever done before. The only way you can do that in a sustained way is through the empowerment of people. And the only way you get empowerment is through high-trust cultures and through the empowerment philosophy that turns bosses into servants and coaches. Leaders are learning that this kind of empowerment, which is what servant leadership represents, is one of the key principles, which based on practice, not talk, will be the deciding point between an organization's enduring success or its eventual extinction. (p. 2)

If you want to lead your team to new heights, you must become a servant first and leader second. Servant leaders lead from behind and within their teams—they work through their teams to achieve the desired results. Unlike other leadership models that rely on antiquated systems of power and a "Do as I say, not as I do" mentality, servant leadership is based on authority and influence. What can you inspire your team to do when you work with them and help them improve? Are your team members better for having served with you? These are the true tests of your tenure as a leader—the mark you make on the organization and the people you serve.

Chapter 5 Review

Leadership and management are not one and the same. Management is oriented toward tasks: planning, budgeting, organizing, and solving problems. A manager's job is to ensure that the team is hitting its numbers and that vital activities are performed. A leader, on the other hand, must be involved with more than the task level and must develop all sides of the team triangle, including the team and the individual.

A leader must do more than ensure a task is completed. A leader must motivate his or her employees to complete the task in the proper manner, in accordance with the organization's guidelines, and in a way that fulfills the purpose of the task. Leadership is about attitude, character, and the ability to influence a team to get things done.

Leadership styles vary dramatically, but leadership is not a personality trait that people are born with. Leadership skills can be developed, nurtured, and improved upon.

Servant Leadership

One of the most effective leadership styles is that of the servant leader. The servant leadership model emphasizes serving followers while leading— working with and through the team to enable them to reach the greatest heights. Servant leaders lead from behind, providing support, encouragement and guidance as the team works through a project or task.

Leading from behind is critical to stimulating innovation and creativity within a team. While the leader is leading from behind, other team members must step up to the plate and take the lead. This rotation of leaders enables team members to develop and explore their strengths as they take their turn at the helm. By leading from behind, servant leaders enable their teams to grow in ways they never could have imagined.

Whereas traditional leaders may rely on power, servant leaders motivate people to serve their purposes using influence. Whereas power can be delegated, bought, sold, and taken away, influence is a lasting power that must be earned.

Servant leaders are all of the following:

- Good listeners
- Empathetic
- Aware
- Able to heal the flaws within members of the team
- Persuasive

- Able to conceptualize
- Capable of foreseeing the potential outcome
- Stewards of their team's needs
- Committed to helping their people grow
- Dedicated to building a sense of community within their teams
- Able to distinguish between their followers' wants and needs

Section 2

The Extraordinary Project Leader

6

Concrete Leadership

The first responsibility of a leader is to define reality.

—Max DePree

Although preceding chapters have made the case for taking the emphasis off of the individual and placing it on the team, one must nonetheless recognize that the project manager is a person who must have a series of prerequisite qualities and abilities in order to be able to develop good leadership and focus on the team.

In order to lead a team effectively, project managers must first hone several essential leadership skills: achieving commitment, embracing accountability, communicating your message, and focusing on results.

Achieving Commitment

Throughout this book, we have emphasized the importance of developing the team's commitment to both the project and the organization. Commitment is not easily gained. Leaders must be able to demonstrate to their teams why they should be committed. One of the best ways leaders can get commitment from their followers is by demonstrating a high level of commitment themselves.

There is a tremendous difference between being involved and being committed. James Hunter suggests the distinction between involvement and commitment is evident at the breakfast table, on a plate of bacon and eggs. The chicken, he says, was involved; the pig was committed.

In order to develop steadfast commitment from your team, you must move beyond simply being involved and become as committed as you can be.

Commitment is defined in onlinedictionary as "the trait of sincere and steadfast fixity of purpose, the act of binding yourself (intellectually or emotionally) to a course of action." Commitment requires that the leader stick with his or her team and project, even during the most challenging of times.

Committing to becoming a better leader, for instance, requires that you work steadfastly on improving yourself for the good of the organization. You commit yourself to excellence, and keep working toward being the best you can be no matter what hurdles you face.

Excellence is a funny thing. People are attracted to and inspired by excellence. People want to be a member of something excellent, to be able to say they had a role in the inevitable success. One need look no further than professional sports to see evidence of the magnetic appeal of excellence. Sports fans select teams to devote themselves to for many reasons: geography, upbringing, or heritage, to name a few. However, many fans choose to support a particular team because the team is the best. The team excels at the sport.

Sports fans are committed to their teams. Even if a winning season drew them in as fans, avid sports fans will stay dedicated even through the toughest times. The Chicago Cubs, for instance, haven't won a World Series in more than 100 years, yet their fans are among the most loyal in all of professional baseball.

Along with that commitment comes ownership. How many times have you seen grown men screaming, "We did it! We did it!" when their team wins a big game? Those men didn't do anything to facilitate the victory, unless the team's fans' ability to sit on the couch and drink beer suddenly became a deciding factor in the score, yet they willingly accept ownership over the team's successes and defeats.

You don't have to encourage fans to stay committed because they want to be part of something excellent, a part of something special. Supporting a professional sports team is one way to achieve that goal. If your team wins the Super Bowl, you can wear a shirt that says "Champion" for a year.

Leaders can harness that level of commitment by helping their teams achieve excellence. Many people have a burning inner desire to be the best they can be; they just often lack the tools to reach their full potential on their own. If you can help your team excel and become the organization's champions, the team members will demonstrate commitment beyond anything you've ever seen.

Developing Personal Rapport

Another excellent way to gain commitment from your team is to make the endeavors personal. A very wise man once said, "People work hard for money and harder for people who will work the hardest for a cause they believe in."

This is excellent advice for any project manager. He or she should understand that a project manager must manage a project, deal with tasks, run meetings, and collect data as well as get the project done on time, on budget, and within scope. But the more important task of the project manager is to encourage and inspire individuals around him or her to be part of the process and participate actively.

As project managers, we spend too much time worrying about the technical forms of project management and too little time dealing with the intangible, personnel side of project management.

If we were to dedicate as much time honing our skills in dealing with people and working with them on a regular basis as we spend looking for new templates and techniques, we would be substantially more successful in the implementation as well as the promotion of project management as an important part of all business.

The problem with dealing with this side of project management is that rapport is difficult to measure. It also takes a long time to see results.

Although the results from "soft leadership skills" take time to become apparent, they will be powerful and much more rewarding than you could ever imagine. Project managers who work long and hard on the personal rapport side of project management not only develop a lasting feeling of accomplishment, but also find lifelong friends and inspire individuals to work harder for the cause.

From time to time, we must spend time reflecting and recognize all of the individuals who have helped us along the way. We must also remember that people do the projects, not the other way around.

Open Your Kimono

Old-fashioned management thinking takes the position that the only motivating factor an employee truly needs is a paycheck. Of course, this line of thought is misguided.

Employees have many different reasons for being part of a corporate work environment; their salary merely ensures that they show up. How a good leader motivates each individual should depend on the individual in question's personal needs and motivations. Motivation, is after all, an individual endeavor. It therefore cannot be a cookie-cutter approach across the entire organization.

One of the best things a leader can do to build commitment within his or her team is an exercise commonly referred to as "opening your kimono." When team members provide feedback for one another using 360-degree interviews or other techniques, leaders can set an example for their team on how to accept constructive feedback and criticism by opening their kimono.

James Hunter relates the story of a project manager who took the concept of opening his kimono to a new level. Upon seeing his results from a peer review, this man called his team in for a meeting. He said, "Man! I got a little feedback here about my moodiness. I had no idea that my moods were having such an impact on you guys. It seems that when I'm walking through this building looking like somebody just shot my dog, that I kind of ruin your day. I'm sorry for that, and do not deny that times have been tough. In fact, my wife and I are going through a divorce. Now that I know how my moods are affecting you guys, though, I'm going to do something about that. My problems are not your burden."

The project manager went on to explain his goal: to come in with a smile on his face, and not let that smile slip into a frown throughout the day. He even had a plan for measuring his progress: he posted a huge chart outside his office door and tracked his mood swings.

The effects on his team were incredible, and his actions sent a message that resonated with the entire organization. Everyone who saw this chart, and learned about the project manager's dedication to his goal, saw that this leader was committed to being the best he could be. He was committed to being a better person.

When people realized that this leader was actively working to improve himself and his game, they were inspired to work on improving themselves and committed themselves to tackling their own flaws.

To open his kimono, the leader must do the following:

- Share the results of the critique with the team.
- Outline his or her action plan for addressing any weaknesses highlighted in the report.
- Begin measuring his or her results.
- Report back to the team on a regular basis about his or her progress.

By opening your kimono and demonstrating to your team that you're willing to work at becoming a better leader, your people will be motivated to work on their weaknesses as well. Remember: project leaders work to "raise their game" so that they can help their followers to do the same. The great test of leadership is whether your team members are better for having followed you. By demonstrating willingness to improve yourself, you team members will be committed to improving their weaknesses as well.

Embracing Accountability

Developing a sense of personal accountability in your followers is absolutely essential to your success as a leader. Team members must be held accountable for their actions, and willing to accept responsibility for those actions. The first step in developing accountability is setting expectations.

Setting Expectations with Your Team

Project managers have many roles and responsibilities in the beginning of a project. One frequently overlooked but crucial task is setting expectations with your team.

These expectations can be in regard to behavior, performance, or both; the important aspect is clearly defining your explicit expectations for each member. Think of setting expectations like setting down the ground rules for a game. Ground rules tell team members what behaviors, attitudes, and results you expect, and what you don't want to see.

These expectations can be outlined in a simple conversation during a meeting or via detailed documentation. Of course, setting expectations is only the beginning. You must hold your employees to the standards you've set along every step of the project, or else the guidelines become meaningless.

If you are expecting something to happen at the end of each month, then that action must occur every month, no matter what. If team members are on vacation, they'll know to make arrangements to have the task completed in their absence—or have the paperwork delivered to you at the appropriate time.

This adherence to expectations helps establish a healthy working environment and allows you to set high standards for the team to live up to. Setting expectations also enables you to take corrective actions when necessary. Consider traffic laws for a moment. A police officer cannot write a driver a speeding ticket if the speed limit was not posted. The expectations—the rules—must be set first.

Remember: set expectations and then achieve them. The results will give your team—and your stakeholders—a much greater comfort level in your position as team leader.

Let the Individual's Actions Dictate Your Level of Involvement

Individuals who continue to meet or exceed the agreed-upon deliverables gain the right to work more autonomously on the project. On the other hand, those individuals who have trouble achieving the results will need you, the project manager, to work more closely with them.

This is not a license to micromanage; instead, try putting more gates and decision points in each task in order to monitor progress from a distance. You'll see red flags as they arise and be able to intercede before a task becomes critical.

There is a fine line between project management and micromanagement, and the project manager must be careful not to cross the line at any time. The line is not static; each individual has different needs. One teammate might feel, rightfully so, that excessive deadlines indicate micromanagement, while another teammate thrives under periodic checkpoints.

The only way to determine whether your actions are having the right effects is to monitor results and have open and honest conversations with each team member. In time, you'll develop a wide knowledge base from your experiences.

Your meetings with each team member may not always be easy. Some team members in need of extra guidance may not feel as if the attention is merited. Despite their protests, even the most experienced individuals should not be left to their own devices. This could spell serious trouble for the project in the long term.

The project manager must check in with all team members on a regular basis—sometimes in a very formal setting for those not achieving their goals. A more informal approach can be used for those who have continually performed and gained the trust of the project manager. However, all team members must understand and respect that it is a project manager's job to monitor the overall project down to the task level to ensure the success of the project and the team.

Accountability is about setting standards, upholding those standards, and helping people become the best they can be day in and day out. One of the best marks of a leader is whether the people they serve are better for having followed—and the only way people will ever improve is if you hold them accountable for their actions.

Discipline and Punishment

The old style of management may have a reputation for being tough, but the reality is that most corporations fail to hold their employees accountable. Many organizations' approach to accountability is better described as avoidance. Organizations lay off, transfer, and fire their problems, rather than addressing issues with personnel as they arise.

Managers sometimes hide their heads in the sand when an employee steps out of line because they don't want to be the bad guy and dish out punishment. They would prefer to just turn their heads the other way and hope that the problem somehow disappears. Unfortunately, when we fail to hold our followers accountable for their actions, we effectively stop them from ever improving, and we serve ourselves and our own interests rather than serving the interests of our teams.

The final element of embracing accountability is discipline. You have a moral obligation to uphold the expectations that have been set and require accountability from every member of your team. Many leaders shy away from the word "discipline" because they associate discipline with punishment. The two terms are related; however, the order in which discipline and punishment are administered makes a world of difference in the final result.

Discipline is a corrective action. Discipline, in itself, does not involve dishing out consequences to an individual for his or her actions. Discipline requires that you identify actions that are not in line with the agreed-upon expectations, and then demonstrate to the offender what the proper action should have been.

Punishment comes after discipline. Once the team member has been reminded of the expectations and guided toward appropriate actions, such individuals must face the consequences and take accountability for their actions if they fail to meet the agreed-upon expectations again in the future.

No one benefits from a lack of accountability; in fact, our tendency to avoid holding employees accountable until the problem gets out of hand actually harms organizations in a number of ways.

For starters, avoiding accountability hurts the employee. If the employee's negative actions aren't addressed, he or she will never have the opportunity to get better. We hide behind a façade of being kind to our employees, when in all actuality we are stunting their growth and encouraging mediocrity to thrive.

Avoiding accountability also hurts other members of the team. When a project manager avoids confronting a team member who isn't pulling his or her weight, that manager breeds frustration in other members of the team. Your team will never be able to achieve excellence when they're pulling dead weight, and their commitment to excellence and the final result will become diminished.

Furthermore, failing to hold team members accountable hurts your credibility. If you constantly espouse excellence, personal improvement, and being the best you can be with your team and then hide in your office when a team member continuously fails to meet the team's expectations, the other team members will doubt the sincerity of your words.

As Ralph Waldo Emerson once said, "Your actions speak so loudly I can't even hear you anymore."

You can't just phone in accountability. If you transfer problem employees to another department without first administering discipline and showing the employee the right way to perform, you've done a disservice to your team and the entire organization—and demonstrated that you are not capable of holding yourself accountable, much less the rest of the team.

You must set the standards by which you will benchmark excellence, and then regularly work to hold team members accountable for any gaps between their actual performance and the set expectations. Notice that I said "regularly" work to hold team members accountable. Timing is essential here.

Communicating Your Message

Share Stories to Inspire and Excite

At the end of one especially long week, I slumped down at my desk. Trying to avoid a confrontation with the mass of e-mails that had accumulated during the day's seemingly endless string of meetings, I reflected on the project consuming my time.

My company was in the midst of an international project to develop an integral piece of software. The project was expected to take the firm to the next level, and all the key pieces had fallen into place: I had the stakeholder's support, the best resources in the company, and a budget that actually made the project possible. Nevertheless, things weren't working, and the project appeared to be headed off the tracks.

For the life of me, I was unable to pinpoint why the project was not going as planned. I had done my homework and checked off the prerequisites for starting a new project. We had hosted a kickoff meeting, complete with everyone's favorite teambuilding exercises. My project leadership exemplified the guidelines laid out by the organization ... but the group assembled had yet to turn into a team. Members did not follow procedure, much less bother to show up for critical meetings. My messages and directions were not being heard, much less implemented, eliminating any chance of a successful outcome. Without increased buy-in from my team, the project was going nowhere.

I felt like Sisyphus, pushing a rock up the hill only to be crushed on its descent back down. Hoping to avoid work for a few minutes, I opened my e-mail and clicked on some personal correspondence. The message was from an old friend with whom I hadn't connected in a long while. While reading his message, I found myself remembering a piece of advice he'd given me years ago, which happened to be some of the best counsel I'd ever received. I was preparing to make a speech in front of 400 people, and being a new and very nervous public speaker, I felt that my chances of delivering a worthy message were slight. As I walked past, he shook my hand and whispered in my ear, "Remember, the best leaders teach by telling stories."

Although that sage advice did little to change the outcome of my speech that night, his words changed my perspective of the world. Upon reflection, I realized that all great leaders use stories to motivate, teach, excite, and make memorable the important things in our lives. Once my eyes opened to this incredible teaching advice, I began spotting examples of teaching stories everywhere. In all major religious texts, the important lessons are told in story form to make them easier to understand and remember. Christians learn about the dangers of succumbing to temptation through the story of Adam and Eve. The tale of Noah and his ark teaches children that faith in God will be rewarded. Confucius explained difficult problems by setting them in a local context. Schoolteachers teach children about historical figures by sharing stories from their lives. Great people of the past did not remain in my head because of the dates they lived or the impact they had on our world, but because I knew stories about them. Few people could describe the origin of Paul Bunyan's historical significance (or even realize that he is a fictional character). Still, most will remember that the mythical giant was a flannel-clad lumberjack with a big red beard who created the Great Lakes as a water source for his beloved companion, Babe the Blue Ox, because of the stories they heard as children.

Once I realized the power of stories as a teaching tool, my whole world seemed to increase in clarity. My father, the archaeologist, had always amazed me with his ability to recruit volunteers. For reasons I could not understand, two or three hundred people decided to fly to the Middle East every summer to labor—without pay—in a hot, sweaty desert for five weeks in a city that had been dead for nearly 2,000 years. Not only could my father convince these volunteers to come and work for free, but he also led them to believe that they should pay their own way. The majority of these archaeo-logical assistants were not the idle rich, either. My father's teams were largely composed of college students and professors with a lot of interest but very little money. Dad's ability to convince people to pay their own way to exca-vate in miserable environments had always been a great mystery to me—but one day I realized that the answer to his successful recruitment was really quite obvious.

My father was a master storyteller who could excite people about the idea of the unknown. He didn't persuade people to join his adventures with promises of discovering great wealth, but rather through the opportu-nity to get a front-row view to what he called "footnotes to history." Dad's digs were nothing like the Indiana Jones movies. In some forty years as an archaeologist, his greatest discoveries included a road system and an ancient religious hall for a long-deceased religion. His teams faced hot, sweaty work with no monetary reward, but they were motivated by another incentive: stories.

While the volunteers labored in the sun, exposing hidden layers of the earth using tiny spoons as tools, my father would talk about how the city was built state-of-the-art with the best of Roman engineering. He would speak the names of individuals from the past that set foot on the walk they were currently excavating and share stories from these long-forgotten lives. His stories were not limited to tales of historical people and places. Dad also told stories about the people in the present and the much more recent past. He would tell volunteers how his teams uncovered the road systems using new technology and ingenious methods discovered in a quest to unveil a city of dreams. He brought the excitement of past digs into the present through his stories and helped people feel as if they were part of something worthwhile and exciting.

Year after year, amateur archaeologists would return to rekindle the excitement and volunteer for more hot and sweaty excavation. These sea-soned workers brought with them their own stories of past excavations. They shared their stories with the new volunteers to convey the fun, hardships, and excitement associated with this once-in-a-lifetime experience.

My father has long since retired from excavating in the Middle East. At eighty-eight years old, he is still convincing people to help him write books about his memories of excavations. His legacy is carried on, almost twenty years later, thanks to new individuals who carry on the traditions and share his old stories along with their new ones. As project managers, we face a similar task as my father. We lead groups of individuals who, although paid, do not report to us and must be treated like volunteers for all intents and purposes.

While reflecting on my old friend's words and considering my current struggles, I realized I wasn't meeting my responsibility as project manager. By becoming so concerned with the mechanics of project management, I had forgotten to treat my team like volunteers and inspire them as best I could.

Sharing stories is a powerful way to ensure that your message is not only heard but also remembered. Rather than simply passing on worn-out corporate slogans and mission statements, you can incorporate a narrative that illustrates the main ideas your team needs to take with them at the end of the day. Stories are a way to exemplify the performance you expect from your followers, inspire them, and excite them about the task ahead—without boring them to death.

When Federal Express first started, the company's motto was "When it absolutely, positively has to get there overnight." During employee training, FedEx often tells their new recruits a story about a letter carrier who arrived at one of the drop boxes only to realize that he had forgotten his key—and it was too late to go back and get it. FedEx boxes are filled with cement and bolted to the ground so that they cannot be knocked over, but this employee wrestled the cement-filled box out of the ground and into his truck to make sure that the packages were delivered on time. Talk about going above and beyond—this employee moved a 500-pound box all by himself just to make sure he got the job done for FedEx's customers. Rather than being punished for forgetting the key, the employee was rewarded for coming up with a way to do his job in difficult circumstances.

By relating this story to new employees, FedEx is able to emphasize the importance of absolutely, positively getting the packages out to customers on time no matter what obstacles may arise. Employees learned that the packages reached their destinations on time—and the driver was not punished. Whether or not the tale is true is almost immaterial, although I hope it is. Either way, narrating the hero's struggle to ensure his company's promise is kept exemplifies to new drivers how seriously FedEx takes its commitment to on-time delivery.

Using Stories as a Memory Aid

The other great thing about stories is that they are easy to remember and easy to repeat. I probably would have had a hard time identifying FedEx's motto before hearing this story, but it's now forever embedded in my brain. In fact, stories are frequently used as a mnemonic device to remember long lists and other large quantities of data.

Review the short grocery list below:

1. Wonder Bread
2. Skim milk
3. Butter
4. White cheddar cheese
5. Sea salt
6. Strawberry
7. Cream cheese icing

Now close the book, wait one minute, and try to write down as many of the list items as possible on a sheet of paper.

How did you do? Were you able to remember all of the items? Now read the silly story below.

> Sally is making grilled cheese. She spreads butter on the Wonder Bread and adds slices of white cheddar cheese. While making her special skim milk and strawberry dessert, the sandwich burns on the stove, so she covers the black spots on the bread with cream cheese icing and sea salt sprinkles.

Once you are finished, close the book and wait for another minute. Now write down as many items as you can remember from the story.

Did you do better this time? Stories are a useful memory aid, and a tool that almost every project manager could use to help better their teams' comprehension and retention of important messages.

Getting Started with Stories

Team members need to be motivated, excited, and focused. Project managers must help everyone understand that there is a process to follow and emphasize the importance of that process. Using stories, project managers can teach their project teams about the expectations and parameters for successful performance.

I keep a small notebook with me at all times full of good ideas and positive actions my team has exhibited. At the beginning of most meetings, I start with a quick story from my notebook about something that I saw during that week, and I recognize that individual by name. The action may be something as small as getting the information to me on time. Recognizing an individual who got me the information on time for the first time breaks the ice—and encourages that individual to keep up the positive practice. Best of all, the more I looked for stories to share, the more I was able to find. From members of my team to senior management, someone was always taking an action I could use to illustrate important points.

Telling stories might set off a chain reaction in those around you. During meetings with stakeholders, I always try to share a story about how an individual from their organization is doing on my team. Interestingly enough, stakeholders then tell me stories about the individual I can take back to share with the rest of the team. Storytelling creates an echo chamber throughout the entire organization. One day, while sorting through a problem on a Gantt chart, I overhead a senior member of the team retelling a story to a new member to help her understand the expectations of the project and why it was so critical to complete certain tasks on time. Good stories spread like wildfire.

By using stories, I was able to turn around the attitude of my wayward team. Stories also made my life easier, because I now have a mechanism to emphasize important teaching moments while also complimenting team members in a very personal way. Once I began sharing stories, the entire project team embraced the practice—and kept it up. Almost ten years later, I stay in contact with the members of that team that once made my life so miserable. The funny and memorable tales from those days still circulate through e-mail and pervade the quiet corners of the local pub when we get together for a brief reunion. Some of the processes we developed and the software we implemented are still working today, which I believe is a testament to how powerful stories can be.

Tips for Telling Better Stories

- Your stories will have more effect if they revolve around other people, rather than yourself. By sharing examples of other people, you can begin using stories to recognize people who have done good work within the project team.

- If you struggle to find stories in the beginning, don't worry. Once you start looking for stories, you will see them everywhere. Furthermore, once you begin telling stories, people will reciprocate and tell you stories as well. You should always remember not only the story, but also the person who told it if you choose to share the tale again within a storytelling group.

- Stories can be used positively or negatively. It is vitally important that you only tell positive stories and reinforce the things that you want to happen. Project managers set expectations, and using stories that reflect negatively on the group will have a multiplying effect.

Develop Your Listening Ability

Communicating your message to others is only half the battle. Great leaders must also be able to receive messages effectively. Listening is much more than hearing what each person says. Hearing is simply the act of receiving an auditory message—hearing has nothing to do with actually processing what the speaker is saying. Listening, on the other hand, goes beyond hearing the speaker's words to include interpreting the meaning of what people are saying with their words and other forms of communication. Leadership communication is about the following:

- Listening
- Facial expressions
- Vocabulary
- Language
- Dialect
- Voice inflection
- Demeanor
- Public speaking
- Body language
- Eye movement and contact
- Appearance
- Response skills

Deciphering Conversation

Most people are not great communicators. Many people say one thing while meaning something else. This could be because of societal rules of what might not be considered appropriate to express, ignorance of vocabulary or language translation, inability to communicate their thoughts or ideas, the limits of vocabulary or translation, emotional considerations, position, the environment, political influence, or response gauging (testing the waters). Therefore, sometimes leaders must "read between the lines" when people talk to them.

Sharing Dialogue

People have a tendency to guess what other people are going to say before they've finished speaking. This restricts what both parties take away from the conversation, because each participant is forming his next thought based on what he thinks the other person is going to say. During this time, no one is actually listening; therefore, much of the information that could have been conveyed during the conversation is lost.

Here's an interesting exercise to make you more aware of any bad habits you may engage in during dialogue with others. You and a partner should begin a conversation with two simple rules:

1. Participants in the conversation will take turns speaking one sentence at a time.
2. Each sentence must begin with the last word of the previous person's sentence.

For example:

Person A: Today I'm going to get coffee.
Person B: Coffee is one of my favorite things to drink in the morning.
Person A: Morning is a time I don't do well with.
Person B: With your help, I can complete this project faster.

While this dialogue activity may seem elementary to anyone with good listening skills, this exercise forces individuals to listen to the other person's entire sentence before formulating their next thought, while also facilitating the exchange of information. Showing your team members how to listen and cooperate in a conversation is a powerful learning tool.

Focusing on Results

Begin with the End in Mind

In the very beginning of a project, the project manager's first task is to be as clear as possible about the outcome of that project. Many sculptors and artisans would claim that "you begin with the end in mind." This is also very true for project management. Michelangelo supposedly would look at a piece of rock and could see the sculpture trying to get out. He would then only cut away the parts of marble holding the sculpture in.

Not all of us can be Michelangelos, but we can envision and describe the project not only to ourselves but also to our teams to confirm understanding and get everyone on the same page. If the project manager goes in with the assumption that he or she understands the desired results without checking with the customer, the project manager has made a mistake before the project has even begun.

Creating agreed-upon results in a project is as vital as agreeing upon the rules before you start playing a game. Beginning with the end in mind can be equated to developing the goal lines of a soccer field before you start playing. If a game of soccer is started without these boundaries, you would just be a bunch of people playing kickball, not soccer.

Setting up those endpoints in a project gives the project manager focus and an ability to track how well he or she is doing over the course of the project, and gives them an opportunity to discuss the boundaries and requirements of a project with the team before the project begins.

Recharge Your Batteries

There is an increasing awareness within many organizations that project management is not only helpful but also vital to the company's operations. As a result, many harried project managers are finding themselves having to coordinate multiple projects simultaneously. Having to deal with competing priorities, individual timelines, and the needs of the varying organizations can put a strain on any leader fast.

Many project managers get overwhelmed by this process and look for a different organization or a different division, believing there's got to be a better way. In reality, most organizations have more projects than they have people or resources. Therefore, creating priorities and understanding the needs of the organization become vital to your own survival.

If you are fighting the urge to run for the hills, you may need to take a step back and remember why you decided to become a project manager in the first place. We refer to this as "recharging your batteries."

When a project wraps up, many leaders jump headfirst into their next assignment. Instead of diving straight into another task, first take a step back to reflect on your accomplishments. Share your feelings of victory by recognizing the accomplishments of the team, and together sit back and revel in a job well done:

- Take a moment to recognize major milestones in the group's progression to a team, and affirm how well the team is working.
- Look at the entire project and the immense amount of work that has been completed.
- Keep lists of funny things that happened during the project, and add milestones to your team's master list of "things to be proud of."
- At the end of the project, read through the lists together and help the team recognize what they have accomplished, not only for the organization but also for themselves.
- Send a personal thank you note—preferably handwritten—to each member of the team recognizing their achievements specifically, as well as offering thanks for making the additional effort.

Other ways of recharging your batteries can include education: try learning new methods and techniques to deal with the same problems. This might involve talking to other project managers, attending a conference, or taking a class.

Engage another leader you respect in a conversation about strategies for managing your workload. Share your experiences with a person in another industry who faces the same challenges, frustrations, and successes that you do.

If time off is not an option, take fifteen minutes each day to read something in the project management field. Although most people argue that vacation is the best way to recharge your batteries, most of us do not have as much vacation we'd like. Finding ways to poke your head up above the weeds once in a while may be as important to your career as taking that long vacation to Tahiti.

Chapter 6 Review

Project leaders must be able to cultivate four main leadership skills to succeed. Leaders must possess the ability to do the following:

1. Achieve commitment.
2. Embrace accountability.
3. Communicate their message.
4. Focus on results.

Achieving Commitment

One of the most challenging tasks a leader faces is developing the team's commitment to the project, the finish line, and the organization as a whole. To cultivate commitment, a leader must be able to do the following:

- Demonstrate why the team should be committed to its goals.
- Exemplify commitment to the team on a personal level.
- Work constantly toward self-improvement.
- Inspire and maintain standards of excellence.
- Help their teams develop "ownership" of the project, task, and overarching mission.
- Develop rapport amongst members of the team.

Embracing Accountability

Achieving commitment is only the first step. Team leaders must also hone a sense of personal accountability in their followers. Team members must be accountable for their actions. They must also be willing to accept responsibility for the consequences of their actions. Leaders can help their teams embrace accountability in a number of ways:

- Setting expectations with the team.
- Varying their level of involvement with team members based on the team members' abilities and needs.
- Understanding the difference between discipline and punishment, and doling out appropriate disciplinary actions at the appropriate time. Timing is essential.

Communicating a Message

Team leaders must have a vision and the passion to move their teams toward that common goal. They also need the skills and tools to communicate their vision to their teams—and to hear the messages that their team has to share.

One of the best ways a project leader can communicate his or her message is by sharing stories. Stories are a powerful way to communicate a message, to reinforce a point, or to help a team remember critical details. Stories can also be used to convey corporate principles, encourage positive behavior, or highlight unacceptable results.

Focusing on Results

The final leadership skill that project leaders must hone is the ability to focus on results and begin with the end in mind. By focusing on results, project leaders have the ability to track their progress and identify potential obstacles or challenges. Focusing on results creates an opportunity to discuss boundaries and requirements and identify what a successful outcome looks like.

7

Dealing with Change

In times of change, learners inherit the Earth while the learned find themselves beautifully equipped to deal with a world that no longer exists.

—Eric Hoffer

Expect Change

Even the best project managers cannot plan for all eventualities. From time to time, our teams will encounter obstacles that they do not know how to handle. The best way to cope with these unexpected occurrences is to recognize that these things happen, that obstacles are an inevitable part of the process.

More importantly, you must demonstrate to the rest of the organization that these sorts of changes, although rare, do occur. This reality should be addressed from the start of the project so that team members can get comfortable with the idea of change and understand that you will help them overcome any obstacles that may occur. While there must be emphasis on the established plan and processes, your team cannot be afraid to deviate from the plan when circumstances necessitate a shift.

Most people do not like change, so having a plan in place to account for change will help team members adjust. Establish a well-documented change control process, familiarize your team with the steps, and keep information available in a place that is easily accessible to all team members.

The way your team copes with these unexpected situations is a reflection on the project's leadership. In times of crisis, the team will immediately look to the project manager for guidance. If the project manager is running around with his or her head cut off, the team will adopt the same response. If the project manager is calm, cool, and collected, the team will reflect that attitude. As a project manager, you must recognize that change is part of life.

Unexpected events will always arise; therefore, you must expect and embrace change. The way you lead your team through times of change is the true indication of your strength as a leader.

Change Happens: Be Prepared!

A project manager's role is to manage the impact of change on a project. You must have a strong structure whereby you can manage the flux that happens within projects. Almost all projects will require the completion of tasks that appear not to have been planned for and that will force a correction to the established plan.

There are two ways to deal with those tasks. One is an ad hoc process, in which anyone can force any decision without going through a structured review process. Although the ad hoc process might work well for some teams, the preferred response is to use a structured and clear review process that funnels all requests through knowledgeable and responsible parties who understand the project, its impact on the organization, and the potential ramifications of the change being proposed.

A change control process can not only make project management easier but also alleviate stress within the organization.

Resistance to Change

Change, in any form, is costly. The cost might be in time, money, headache, heartache, or a number of other commodities. No wonder why most people resist change when new developments come their way.

Bringing down people's resistance to change is itself a challenge. The key to meeting this challenge is to point out the positive attributes of the change while still recognizing the worries that your team members might have.

For example, suppose that the team is bringing on a new member. This new member has done well on other projects, but worked with teams that had different approaches to many tasks. The team is worried that this might upset the team dynamics, and they are unsure about how much activity they can entrust to the new person.

Make a point of reviewing the new member's qualifications and past history of results. Tell the team that you trust this new person, but that you understand why everyone would want to get to know him or her first. Be honest and say that you recognize that there will be a transition period, and that you are doing everything you can to make sure the transition happens smoothly.

Ultimately, team members will follow your example. If you can embrace change and adapt deftly, they will follow suit. If you do nothing, or if you remain overly critical, your team members will be less likely to embrace change.

Conflict Resolution

Inevitably, change also brings conflict. There are two types of conflict: productive ideological conflict, and nonproductive conflict. The first type of conflict occurs because people have a genuine difference of opinion, and debate ensues in an honest effort to reach the truth (or agree on a plan of action). The second type of conflict includes many different motives—politics, pride, opportunism, the need to vent, or just basic annoyance—but they all share the common trait of being unproductive.

The first step in conflict resolution, then, is figuring out what kind of conflict you have on your hands. Productive ideological debates should not be prevented, although you can provide tools to make sure they run smoothly and do not devolve into unproductive debate. Unproductive debate should be either prevented or cut off when it occurs.

Unproductive debate can be identified by one or more of the following qualities:

- Personal attacks
- Lack of focus on actual policies or actions taken
- Little mention of team metrics, standards, or policies
- Returning to old or resolved issues
- Mentioning historical conflicts or personal details
- Mentioning the same ideas or plans again and again, with little revision or explanation

When you have an unproductive conflict on your hands, you will need to find ways of ending or diffusing the conflict.

Sometimes diffusing a conflict can be done easily with just a word from the team leader, or a gentle redirection to more pressing issues. But sometimes conflicts become challenges because one or more obstacles prevents the conflict from dying. These might include the following:

- *Information obstacles*: lack of facts, diverse perspectives, untimely information
- *Motivation obstacles*: poor corporate culture, poor leadership, uninspiring activities, lack of clear goals, lack of trust
- *Organization obstacles*: poor planning, vague steps or activities, unclear delegation of responsibilities, lack of metrics (or poor metrics)
- *Activity obstacles*: physical constraints, unrealistic deadlines, supply problems
- *Personal obstacles*: personality conflicts, differing expectations (or unreal expectations), lack of appropriate skills, lack of commitment

If a conflict won't die, or just keeps resurfacing, try to see if there is an enduring obstacle in the way, then take steps to remove the obstacle.

Productive conflict is different from unproductive conflict. Indeed, productive conflict should be fostered and encouraged. Productive conflict gives your team members a sense of purpose and an understandable narrative in which to fit key decisions. A good productive conflict can also enliven meetings and spotlight issues before those issues become problems.

How can you encourage productive conflict and prevent it from devolving into unproductive conflict? First, you should have a standard set of policies or "rules of engagement" for conflict and debate. Your policies will reflect your team, your organization, your goals, and the personalities of your team members, of course—but there are some common elements I recommend for all policies:

1. You don't have to be polite, but you do have to be nice and play fair—no personal attacks or the like.
2. Try to stick to one topic at a time.
3. Once someone is talking, give him or her the time and attention due.
4. Try to keep your own points short and sweet. End by reminding us of your main point.
5. Feel free to interrupt with questions.
6. You can walk away disagreeing, but no one walks away angry.

You should also encourage conflicts that you feel are productive while downplaying ones that devolve. For example, if two team members disagree about a product rollout, feel free to step in and say something to the effect of the following:

- "This is a good topic to debate. Go on."
- "Wait, so you disagree. Let's see why."
- "OK, you are having an argument. That's fine—let's hash this out."

This might sound hokey at first, but you'll be surprised at what happens when you give your team members permission to debate (in the correct way, of course).

Aside from encouraging healthy debate that will benefit the team, as a project manager you should stay largely disengaged from conflict when you can. Your role as leader is to act as a mediator and as the occasional tiebreaker. If you must get involved, consider inviting a neutral third party to moderate the conflict.

Coaching and Mentoring

Coaching (or mentoring) is one of the best ways to deal with change. By taking a proactive approach and coaching individuals before, during, and after a change takes place, you can develop greater influence as a leader and continue to retain the highly skilled individuals on your team. Thus a good project leader is, above all, a good coach and mentor.

Project leaders should be able to interact with their teams on many levels and help team members to understand what needs to be done in a positive, affirming way (rather than in a dictatorial or commanding way). The main problem with the command style of Industrial Age management is that managers often step in to solve a problem, but then do not develop team members in a way that predicts or prevents problems in the future.

Take the example of Robert, a manager for a small advertising firm. Robert's firm had purchased a smaller competitor with an excellent customer list and several star employees. Unfortunately, this excellent customer list contained several demanding, high-profile customers, and the new employees had to be integrated into the existing company and way of doing things.

Under the old style of management, Robert would have laid down the company's policies and asked his new star employees to conform to those standards. The customers would have to be reeducated about the process of doing business with Robert's firm, and most of them would likely have tested the waters with only small projects at first.

Instead of this scenario, Robert decided to take a different approach. He gave his new employees the freedom to do whatever needed to be done in order to retain their previous customers. As they did so, Robert would pay special attention to "problem" clients who were either withholding business or threatening to switch to another firm. Before his employees could get frustrated with the situation, Robert would take them aside and politely offer lessons he had learned from his years of experience with the company. The goal, he said, was to have each team run itself with as little input from him as possible.

The efficiency of Robert's approach was soon apparent. Whatever methods were working at the time were kept in place, and clients were not bowled over by the internal changes. Those methods that were not working, however, were addressed, and employees were slowly reeducated.

Soon, Robert's role was less that of a manager and more that of a coach. He would even provide miniature seminars about established advertising techniques and client relations. After about a year, the other managers noticed something: many of the teams that Robert was coaching were better able to deal with high-demand clients, and were much more flexible in terms of internal changes within the company.

There is an old proverb: "Give a man a fish, and he eats for a day. Teach a man to fish, and he eats for a lifetime." This was Robert's motto as well. Instead of stepping in and solving problems one after the other, Robert gave his teams the basic tools they needed to deal with change themselves.

The ability to mentor teams is one of those telltale skills of a modern project manager. The mentoring process can be captured in the five steps of the COACH:

1. Correctly identify problems and challenges.
2. Open a dialogue about those problems with the team.
3. Assess skill areas where the team might be weak.
4. Convey the information that the team needs to address the immediate problem.
5. Hone the skills needed to address similar problems and challenges that might arise in the future.

For more information on how to be a coach and mentor, please see our companion website, www.ManagementToLeadership.com.

Being a mentor and following these steps can be a much more demanding task than one would imagine. There are three crucial attributes of a successful team mentor, attributes that you will need to hone before and while you are coaching:

1. *Expertise.* The information given by a mentor must be tested and credible, and must be relevant to the actual challenges that a team faces.
2. *Clarity.* The mentor must have clarity with regard to the organization's overarching goals, the team's mission, and the problems at hand. A mentor must also be able to communicate their expertise to the team effectively and train team members in the skills necessary to succeed.
3. *Personality.* Mentors walk a fine line. A knowledgeable mentor can still come off as arrogant or meddlesome...but, then again, a mentor that is too "hands off" or friendly can come across as insubstantial or, worse, useless. Mentors need to provide the information and skills necessary for solving concrete problems while avoiding the "Mr. Professor" mentality.

Each company and each team will differ in their reactions to mentors and their personalities. Some groups are perfectly happy to receive lectures with vital information from an expert in their field. Others will see this approach as condescending. Like everything else, striking the correct balance will take some trial and error on the part of the project leader.

Mentoring is great in principle, but how does mentoring actually influence change? First, a good mentor will convey to the team the feeling that no problem is insurmountable. Team members can react to challenges with worry, skepticism, and even panic—but the presence of a coach or mentor can relieve team members by letting them know that there is someone who has seen this kind of challenge before, and that there is someone with a possible solution.

Second, mentoring helps to establish long-term relationships with team members. Rather than a figurehead, the project manager will be seen as someone who cares about employee development and training. This reputation will, in turn, increase employee loyalty and long-term value.

Third, mentoring is a regular, predictable way to give team members new skills. Instead of the project leader anticipating and reacting to changes, team members can be taught to explore trends, expect changes, and modify their efforts in order to deal with change. If there is a good mentor behind a team, one rarely sees the mentor getting his or her "hands dirty," because the team has become an efficient engine for dealing with change.

In the past year alone, businesses have spent billions of dollars on outside coaches and consultants. There is certainly no reason why a good project manager within a company cannot become an inside coach. Indeed, coaching from the inside is likely to be more effective, since the project manager is already intimately familiar with the company, the team, and the product or service. If every manager became a coach as well, the effect would be equal to the hiring of twice as many outside coaches for an entire year. To put this another way, if every manager took an interest and coached just one individual in the organization not only would the company save millions but also morale and productivity would be improved because of the focus and interest that manager took in another employee.

Chapter 7 Review

Change is an inevitable fact of life in any organization or project. Project leaders must learn to recognize that unexpected circumstances are going to arise and learn to cope with these changes as effectively as possible. The way a project leader guides a team through times of change is the true measure of his or her leadership skills. Change often breeds resistance, conflict, and unease within teams. Project leaders must learn to confront these negative behaviors and teach team members to channel these emotions and behaviors in a productive way.

One of the best ways a project leader can help his team learn to deal with changes is by taking a proactive approach and coaching or mentoring team members. By developing a coaching and mentoring system, project leaders can give their teams the tools they need to handle change. The coaching and mentoring process is captured in the acronym COACH.

- Correctly identify problems and challenges.
- Open a dialogue about the problems.
- Assess skill areas in need of development.
- Convey the information needed to address the problem.
- Hone the skills necessary to address the problem.

If team leaders are able to coach and mentor their teams through times of change, the teams will learn to better cope with change and be better prepared for unexpected circumstances in the future.

8

Leadership Sideways

> The only test of leadership is that somebody follows.
>
> **—Robert K. Greenleaf**

Interacting with Stakeholders

Successful project managers must deal with stakeholders at some point in their projects. These are individuals who are significant to the project, but who might not have direct influence on the project itself. For example, if the purpose of the project is to compile a new customer service database, the customer service representatives inputting the data might not have much say in the project but could make or break your success with the project. Therefore, we'd consider them stakeholders.

Stakeholders might also include people who have invested in the project. This investment could be money, but investment can also be in time, political clout, or resources.

Savvy project managers develop their rapport with stakeholders as much as with their own team. Knowing the crucial attributes of stakeholders is not as critical as it is with team members, but you will want to know (1) what the stakeholders have invested, (2) what their goals and aims are, (3) how they deal with conflict, and (4) what sorts of questions they will ask of you. For a good stakeholder management template, please see our companion website, www.ManagementToLeadership.com.

Since the project manager is the mouthpiece of the team and the cheerleader for the team's project, he or she will, at times, need to communicate with stakeholders on the progress of the project (or report on problems that the project has encountered). There is a balance to strike here: one should always be honest and forthright, but one should also put the most positive light on one's project, one's team, and one's team members.

The best advice for dealing with stakeholders is to underpromise and overdeliver. If you can continuously beat expectations, you open the doors of opportunity to good rapport in the future—and you also garner a reputation for getting things done.

Exciting Team Members

I want to share with you the story of Kyle, a sales manager at a small printing company. Kyle was one of those managers who is easy to like at first: he had a broad smile that he used often, was excited about his job and his products, and could relate to any customer he spoke to—even the ones with off-the-wall hobbies or deficient social skills. Kyle obviously knew the printing industry inside and out, and was eager to share the things he had learned, both about printing and about the sales process, with anyone who would listen.

I spoke privately with most of the employees who worked under Kyle, and they all gave me roughly the same picture:

They hated Kyle—and they felt that Kyle was an awful manager.

In fact, the entire sales department of this print shop had turned over not once but almost twice in one year—about 150% employee turnover. Employees showed little to no sales growth in the time they were there, and some even lost customers they had brought over from previous positions with the competition. Department morale was consistently low.

How could someone with a seeming overabundance of friendliness, charm, and expertise end up being one of the worst managers these people had ever worked for? When I talked with Kyle's employees, here are some of the complaints they cited as reasons for Kyle's incompetence as a manager:

- Problems were listened to, but never addressed.
- Too much office politics.
- Inconsistent policies.
- No clear expectations, or contradictory expectations.
- People's particular talents were ignored.
- Projects were micromanaged.

Anyone who took the time to look could see all of these behaviors in what Kyle did. He often assumed that he knew more than his sales team, and so problems were often dismissed as personal flaws, or simply as not being worth his time. This assumption also meant that Kyle often tried to micromanage projects and sales calls. Kyle's constant appearance over his employees' shoulders communicated a basic distrust in their competence, drained away the enthusiasm of his salespeople, and prevented them from learning on their own. And because Kyle thought he knew the right way to make a sale, he often missed the particular talents that each salesperson brought to the team, and quite often mismatched salespeople and potential customers.

Kyle was also often at odds with the management of the company, which led to inconsistent policies and conflicting sales goals.

These problems are all too common; just about every book on management cites them as "demotivators" and "roadblocks" to developing productive employees. The question is, how can these offending practices be avoided?

The first key to avoiding motivation problems is to remember the following:

Anybody can stay motivated about anything for about five minutes ...
Getting a team to stay motivated is the tricky part.

Employees are naturally motivated when there is an interesting project to work on—especially if the project pays well. But the initial enthusiasm for a project can erode with time. Distractions, fatigue, disorganization, problems, and panic all wear away at the initial enthusiasm until the quality of work suffers. The trick, then, is to keep employees motivated, even during the worst of times.

Employees are naturally excited by interesting and challenging work, yet work is not the only driver of motivation. People are also excited by the following:

- Creativity
- Independence
- Economics
- Status
- Service
- Work conditions
- Academics
- Collegiality
- Security

There are a hundred strategies—from forming a cohesive corporate culture to little tricks of the trade—for keeping the people on your team motivated. These are covered in hundreds of books and Internet articles, so I won't bore you with the details here.

Rather, I will share with you the three tricks of motivation that managers (and management educators) often forget.

Assume You Are Wrong—about Everything

I once worked for an international software company. The head salesperson frequently had to deal with upset clients, and he excelled at mitigating delicate situations.

When faced with one of these less-than-thrilled clients, he would sit down with the client and open by saying, "OK, it's all my fault. Everything! Everything that has gone wrong is my fault, and I accept that. Now, let's sit down and figure out how we're going to fix this situation for you."

By immediately accepting responsibility for his actions and ending the blame game before it began, he was almost always able to instantly transform the situation from difficult to much more placid—and achieved better results.

Assume that you are wrong about everything: your business, your fellow teammates, your boss, and even yourself. This will force you to ask more questions, consider new strategies, and challenge assumptions that might be hindering your business.

Assuming that you are wrong will also help you to listen to the people on your team. When we assume that we know our teammates and the challenges they face, we tend to listen, but not really hear. Listen to each teammate as if every problem is a new one.

Incentives Can Actually Be Disincentives: Use Challenges Instead

From grade school onward, we are trained to do things because we get rewarded—and so most managers use incentives to motivate their employees to work. Ultimately, this strategy can be counterproductive.

A growing body of research is showing that incentives only give temporary motivation and encourage sloppy work. Researcher Alfie Kohn forcefully argues against incentives in his book *Punished by Rewards: The Trouble with Gold Star, Incentive Plans, A's, and Other Bribes.* He emphasizes that employees are naturally motivated by work that is interesting and challenging. The best thing you can do is give them an interesting project and let them run with it.

Of course, there are exceptions to every rule. Some team members might be motivated by incentives; a good leader should be able to identify these team members and adjust for them accordingly.

Motivate Yourself First: Others Will Catch On

Too much emphasis is placed on motivating employees, but often it is management that can't wait to get out of the office and away from the grind.

Passion is contagious. The best way to get employees excited is to be excited yourself. Research your industry in your spare time. Talk about your work with anyone who will listen. Describe the outcome you want from each scenario in vivid detail. If you are genuinely passionate about the outcome you want, your employees will see the outcome as something valuable in itself.

Leading within the Company

> "If there is no community for you, young man—young man, make it yourself."
>
> **—Paul Goodman**

Not everyone can lead an organization. Furthermore, not everyone can (or is given the authority) to lead a team. In these cases, you must empower yourself to become a leader. Although you might not have an assigned team to support you, you can use your influence to develop a network of individuals to support you in your endeavors.

How does one develop such a network? To be honest, there is no sure-fire, easy way to build a support network. The process takes time and foresight. Influence and credibility can only be built slowly, with great attention to detail.

With that said, there are a number of habits that you can cultivate that will eventually lead to the aforementioned influence and credibility.

Take an Interest in the People around You

This habit is easy for some, but for others taking an interest will require some conscious habit forming. The core part of taking an interest in others is listening. Whenever you can, find out more about others. What are their likes and dislikes? What are their complaints, and what gets them excited? What challenges are they facing?

You need not go on fact-finding missions to discover this information. Just be a friend to those around you. Develop your sense of humor. Reach out and take an interest, even if only briefly. People tend to remember those who took a moment out of their day just to catch up.

Harness Mutual Respect

Always treat others with honesty and respect (even if they are not so deserving of that respect!). Try to see the good in people and appreciate your team members' abilities, rather than focusing on their failings. Be open with your information and your time, and be stingy with your criticisms.

Listen to Others' Problems

If you are sincerely taking an interest in others, you will naturally find yourself in conversation where problems and challenges are discussed. Listen to these issues with an open mind. Even if the person you are communicating with is not a member of your team, you can learn a lot from a good old-fashioned gripe. Besides learning, listening to a problem opens the opportunity to provide a solution.

Offer to Solve Others' Problems

Good project managers are also good problem solvers. But just because you are not managing a project does not mean that you cannot be a successful problem solver.

If you have been listening to the problems of others with an open mind, there will come a time when a problem presents itself that you can fix, or at least help ameliorate. The best way to be perceived as a leader is to first garner a reputation as a problem solver.

Avoid Gossip and Office Politics

Gossip does not solve problems—more often than not, gossip creates problems. Rise above the fray. If you have a reputation for gossip, no one will share their problems or challenges with you, which will destroy your ability to become a problem solver. Office politics can also destroy trust and distract people from the true mission at hand. Curb the temptation to engage in such games.

Think "Win-Win"

Under the old rules of management, different team leaders were often competing with each other for many things: the best team members, limited budgets and resources, overlapping client lists, and so on. However, the wisest leaders don't think in terms of "win or lose"—they think in terms of "win-win."

Every challenge can be solved in innumerably different ways. Some ways of meeting a challenge are less savory than others (one response to any business challenge would be to fire everyone and close shop—but that is hardly a solution worth considering!). A little bit of deep digging is sometimes necessary to find the solution where everyone wins.

Take the example of an employee (we'll call him Joe) concerned with "going green" at the print shop he worked at. Management was dead-set against any environmental efforts because of the time, money, and distraction that such efforts would cost. Instead of butting heads with the management, Joe found some simple ways to cut waste.

For example, Joe noticed that a lot of waste paper was generated when large sheets were cut down to size after being printed. There was no way to prevent the waste, but Joe felt bad that the shop was producing box loads of this waste paper every day.

Joe also knew, however, that the owner was spending large amounts of money on costly packing materials. This is when he came to a realization: why not wad up the strips of waste paper and use that as packing material? The strips of paper were free, because the shop generated them as waste every day. And the crinkled paper was as good a packing material as the packing peanuts the shop was buying from its supplier.

Joe presented the idea to management initially as a money-saving trick. Once the "leadership" was on board with the idea, Joe also pointed out that this would be a form of recycling, and that the process would be a great selling point to new clients. He offered to help the sales team in working the environmental angle into their sales literature. Total cost to management: $0 (management actually saved money). Total cost to the sales team: $0 (any reprints of sales materials were done in house). Total benefit to the environment: that many more trees saved. Once this became a success, the group started seeing other possibilities for scrap paper, including making notepads and scratch pads—items that could be sold to their clients.

This story is just one small example, but the idea should be clear: stop thinking of all innovation as a struggle against the status quo. A true leader tries to find a solution that is "win-win" for all involved. Granted, this often takes time and creativity. But the reputation that one gets for finding the win-win situation is unparalleled in the management world. People who think win-win tend to attract the best employees, expand their circle of influence, and make other people more likely to think the same way as well.

Notice that all of these steps can be done without a team, without a title, and without a project to manage. Indeed, these habits are more likely to signal a true leader—someone who can lead without the need for the external trappings of leadership. Few employees distinguish themselves with practices such as these. By cultivating a talent for "sideways" leadership, you can signal to the decision makers in your organization that you are ready and able to lead.

Chapter 8 Review

Leaders must be capable of communicating with people on every level of the project, from individual team members to their immediate supervisors, to the project's stakeholders. Each of these groups requires specific attention from the project leader, and the project leader's ability to communicate with each group will determine his or her success within the organization.

Stakeholders

Dealing with stakeholders is one of the biggest responsibilities of any project leader. When interacting with stakeholders, the project leader must be the team's public relations specialist and its biggest supporter. The leader needs to be able to communicate the progress made on the project, and report on any problems the team is encountering.

The project leader must strike a balance when interacting with stakeholders. On one hand, they must be able to communicate in an honest and forthright manner. On the other hand, they must portray the project and the team in the most positive light possible.

To get the best results from interactions with stakeholders, project leaders must learn to underpromise and overdeliver. This enables the team to not only meet the stakeholders' expectations but also exceed them.

Team Members

Project leaders must be able to excite team members about their projects if they want to deliver good news to the project stakeholders. Employees must be motivated, inspired, and cajoled into performing at their highest levels. One of the easiest ways to excite team members is to give them interesting work—more specifically, work that is interesting to the individual undertaking the task. Another great way to excite team members is to motivate them with custom incentives. When project managers can determine what truly motivates an individual to succeed, they can light the fire under that individual using the perfect incentive as a spark.

Leading within the Team

Not all leaders are actual "leaders," at least not according to the titles printed on their business cards. Of course, true leaders do not need an official title to lead. Anyone can hone their leadership skills and develop a reputation as a leader without being designated as the boss.

9

Leading the Next Generation

> In the transmission of human culture, people always attempt to replicate, to pass on to the next generation the skills and values.
>
> **—Gregory Bateson**

For project leaders, an emerging generation of highly talented and efficient workers is posing new challenges. They are called "Generation Y," and they make up an increasing percentage of the workforce. Generation Y consists of people born between the years 1980 and 1995. This generation has specific needs, characteristics, and skills that are unlike those of past generations.

Generation Y is about 60 million strong, which is the largest workforce since the Baby Boomers. They are just beginning careers and working with coworkers and leaders of other generations. They currently make up about 22 percent of the workforce, but the number continues to grow. Over the next five years, 10 million more are expected to join the job market. Some employers are experiencing tension between employees of different generations. According to some surveys, about 70 percent of older employees underestimate the abilities of their younger counterparts. About half of the younger employees feel the same way about their older coworkers. Because Generation Y'ers have such unique traits, it is essential for project leaders to learn how to deal with some of these challenges.

The young people of Generation Y were greatly affected by the political and social events of their time. After the Vietnam War, many Baby Boomers felt skepticism and disillusionment toward the government. Likewise, September 11, 2001, changed many young people's attitudes toward politics. They became more patriotic and socially aware. They became increasingly involved in political discussions and committees. They were reminded that life is short and that work is not the most important part of life.

One key attribute of Generation Y'ers is that they grew up with access to technology. Conversations through e-mail, instant messaging, and text messages are just as real to them as face-to-face conversations. A virtual relationship is just a natural extension of their personal experiences. A survey of college students revealed that about 97 percent own a computer and 94 percent own a cellular phone. About a third of these students use the Internet as their primary source of news. With numbers like these, it is no wonder that Generation Y is suitable for a diverse, fast-paced, and global work environment. They understand how interconnected the world is and are capable of dealing with current global issues.

Characteristics of Generation Y

One characteristic of Generation Y is the need to balance work with one's personal life. Unlike the Baby Boomers, who place a high value on their professional careers, those in Generation Y expect their jobs to be flexible and accommodating to their personal obligations. They are looking for telecommuting options, time management, and an understanding with regard to personal commitments. They see work as a means to an end. This is because many Generation Y'ers were raised with a list of activities. Some had soccer practice, Boy or Girl Scouts, music lessons, and piano lessons all while maintaining high grades. They think they know how to multitask and how to manage their time. (Multitasking is a myth that creates a drain on workplace productivity. Please see Chapter 14 for more in-depth discussion.) They are not content with jobs that encroach too much on their personal life.

Because Generation Y'ers need to balance their work with their personal lives, they dislike rigid schedules and the traditional nine-to-five workdays. These young employees place less of an emphasis on when the work gets finished, so long as it does get finished. The focus is on productivity. So long as the employee is efficient and putting out great numbers, he or she feels there should be no problem with leaving early on a sunny Friday afternoon.

Technology has also influenced this aversion to working a standard day. Those in Generation Y grew up with technology readily available. Life takes place 24/7, so Generation Y'ers expect the same for their jobs. The line between work and personal life is blurring. Although an older employee would never take a personal call at work, a Generation Y'er would have no problem answering a call from a boyfriend or girlfriend at the office. At the same time, they are more likely to make a work phone call outside of the standard work hours. If a project leader asks an employee to come to work at a specific time, a Generation Y'er would expect face time during the entire session and a discussion that would have been impossible over the Internet. An inflexible work schedule is a guaranteed way to lose Generation Y employees.

Young adults from Generation Y prefer to work with friends and are very conscious of group success. Since they grew up with technology, Generation Y'ers have had their family and friends available to them at any moment in the day. They can refer to the advice and opinions of others in an instant. Therefore, Generation Y'ers were bred with a natural inclination toward group work. They are much more satisfied working in teams, which is a characteristic that project leaders can capitalize on. Generation Y also expects the work environment to be fair and diverse. Young people refuse to let others fall behind and will use their collective power if they feel that a coworker is being treated unfairly.

While Baby Boomers may have remained loyal to a company for loyalty's sake, Generation Y'ers do not expect to stay in their positions for long. They have become skeptical of employee loyalty, especially after scandals like Enron and Arthur Andersen. They have high expectations of their employers. Many companies have struggled to keep their young employees. Generations Y'ers will leave a job very quickly if it is boring, meaningless, or too demanding on their personal time. The average career is now about two years, with many Generation Y'ers leaving a job after only about eighteen months. Project leaders must work to keep the jobs interesting and challenging for their younger employees.

Generation Y'ers are willing to leave a position in order to improve the quality of their lives. Recent surveys indicate that Generation Y'ers are more loyal to their lifestyles than to their jobs. If young employees are not making enough money or no longer find their work intellectually stimulating, they will simply leave. One major reason for this is the increase in student loans. If they see jobs that pay even slightly higher, Generation Y'ers will quickly leave their current jobs. This can be very frustrating for employers. They spend time and money hiring and training new employees only to find that they cannot retain them. As a result, companies are forced to come up with new and creative ways to attract and retain younger employees. For example, some offer perks like a one-month sabbatical after five years of work. Such time off would allow a Generation Y'er to explore other areas of interest such as volunteer work or time with family and friends.

One misconception of Generation Y'ers is that they are inattentive and unfocused. The common image of a Generation Y'er is one of a young professional staring at a computer for hours on end. Some employers may not view this as hard work. However, Generation Y'ers are actually very goal-oriented and efficient. They grew up with information available through the touch of a button. These Generation Y'ers were raised by highly involved parents who set multiple lofty goals for them, so they set high goals for themselves. As a project leader, you must make clear to the Generation Y'ers what their goals are and why these goals are important. Even better, provide two types of goals for the younger employee. One type should focus on producing high numbers. This goal measures the output of their work. The second goal should emphasize how their work contributes to something greater than just making money. This goal is geared more toward how the Generation Y'er's work benefits the progress of the company overall. With an understanding and appreciation of these goals, the Generation Y'er will stay focused on achieving them.

Let us look at some of the negative assumptions of Generation Y, especially regarding technology. Some see Generation Y'ers as lacking in problem-solving abilities and independent thought. This view assumes that young employees are overly reliant on technology. The solution to this problem is to challenge employees to solve problems on their own. Managers should be mentors for them, modeling how to deal with new types of problems. The second negative view of Generation Y'ers is that they lack social skills in face-to-face interactions. They prefer e-mailing, texting, and phone conversations instead. This may lead to misunderstandings with older coworkers or clients. Project leaders should provide clear expectations about when face-to-face interactions are expected.

For the most part, Generation Y'ers' technological skills are great assets for a company. If harnessed correctly, these skills can greatly benefit a company by bringing the organization into the modern era, so how can a company harness these technological skills?

One way is to provide access to the most up-to-date gadgets. Generation Y'ers want to use the fastest tools on the market. This will allow them to be as productive as possible. Another way to cater to these technological skills is to allow freedom with the use of technology. Project leaders should be honest about regulations regarding websites like Facebook and MySpace. For the most part, access to these websites will not deter employees from working hard. In fact, the momentary relief of checking their e-mail or Facebook account may help Generation Y'ers be more effective at work.

Within the last thirty years, a technological revolution has occurred. The Baby Boom generation watched as the Internet was created and grew into an essential part of everyday life. Cellular phones, cameras, MP3 players, and computers are constantly becoming smaller and faster. Baby Boomers often feel overwhelmed by these ever-changing technologies. They also tend to feel embarrassed by their own inabilities to keep up with the times. On the other hand, Generation Y grew up right in the center of this revolution. Generation Y'ers are great assets to any company that needs IT workers. They can provide computer assistance and can install the most up-to-date programs. In doing so, they can aid their older counterparts in becoming as efficient as possible.

The technological revolution has also changed how people network. For Generation Y'ers, networking is a natural ability. They are accustomed to using such websites as Facebook, MySpace, and Twitter in order to make and build relationships. Knowing how to utilize these websites can greatly benefit a company. This type of networking allows employees to create virtual teams and to collaborate with others via the Internet. Furthermore, many companies are learning to use these networking websites to attract customers and to increase sales. Employees in Generation Y are capable of utilizing these social-networking websites to their highest potential.

The Project Leader's Role in Integrating Generation Y into the Workplace

Employers need to adjust to the distinct skill set of Generation Y. This involves investing in management training to accommodate for the increase in Generation Y workers. Both project leaders and Generation Y employees need to be trained on how to deal with workers of other generations. Even though this process can be costly, the investment will do more to motivate and reward Generation Y employees.

Listen to the Employees' Needs

An organization must make employees feel valued. Project leaders should listen to the needs and goals of Generation Y employees. One famous company, Deloitte, faced a difficult problem as a result of not listening to the needs of young employees. In 2004, a manager assigned work to some young employees over the weekend. When the employees asked to reschedule the work, as they had made previous plans, the manager was quite upset. This was a recurring problem for the company and created friction between the older and younger employees. Deloitte was eventually removed from *Fortune*'s list of the "100 Best Companies to Work For" as a result.

The national director ultimately decided to train the managers instead of dealing with the younger staff. Afterward, the company no longer faced such problems and was put back on *Fortune*'s list. The point of this story is not to make Generation Y'ers sound demanding. Instead, I am attempting to illustrate how companies can listen and respond to the needs of their younger employees.

Demonstrate That Input Is Valued

Generation Y'ers are looking to have their ideas heard and to be given credit for these ideas. As children, they were encouraged to state their opinions and to question authority. They have watched friends start successful businesses and become self-employed at early ages. People from Generation Y will not respect a leader simply out of respect for power. The best strategy for any leader is to listen to the Generation Y employees. The old way of commanding and expecting loyalty no longer exists. Leaders can gain respect by listening to and getting to know the employees.

Develop Trust between Employer and Employee

Generation Y has an innate distrust for authority. Managers need to be open and honest with employees. Some companies are investing in coaching and mentoring programs in order to deal with new employees. Even though such programs require the use of valuable resources, they can improve communication between employers and employees. They make Generation Y employees feel appreciated and comfortable in the work environment.

Encourage Innovation

Generation Y'ers enjoy providing new ideas and seeing those ideas followed through. They appreciate the freedom and independence of running new projects. If Generation Y'ers feel unable to express their opinions, they will quickly choose a new workplace where they have such freedom.

Provide Opportunities within the Organization

Generation Y'ers look for opportunities to grow and to challenge themselves. They want the possibility of future promotions. I know of many stories in which a Generation Y'er was given a raise without being given a promotion, which can frustrate the employee. The raise acknowledges their hard work but does not provide the opportunity to challenge them further. It might even cause the employee to leave the company sooner than if he or she had not been given the raise.

Involve Employees in Team Formation

Generation Y'ers want to work in teams that work well together. Some young employees become very irritated when placed in a team that does not push them to learn new things, even if they are making a high salary. When on a team, Generation Y'ers have certain networking preferences. Organizations should allow employees to network with teammates in any way they choose.

Consider Alternate Working Arrangements

With the Internet as a resource, Generation Y'ers want companies that allow alternate working arrangements. For example, the standard workday is quickly being replaced by telecommuting options. Employees are able to work from home or from local coffeehouses by using the Internet. Video conference calls, instant messaging, and virtual networks make such arrangements possible. These arrangements require a new type of virtual managing, which can be difficult for older managers. However, companies will benefit by marketing to people worldwide and by providing more comfortable work environments for Generation Y employees.

Require Them to Pay Their Dues

When Generation Y'ers are applying for a job, managers should be clear that rewards come only after employees put in their dues. Generation Y'ers were raised with a lot of praise and encouragement from adults. They want rewards quickly. Many Generation Y'ers were sheltered from criticism and do not always handle failure well. They should be prepared for a tough apprenticeship first. They need to get through the demanding work and difficult criticism before they can be rewarded.

Allow This Generation to Manage People and the Last Generation to Manage Tasks

A Generation Y leader needs to inspire employees to be leaders. Generation Y'ers are great at building relationships and managing people. As they are better at dealing with people, Generation Y'ers are more capable of managing people than tasks. A smart leader will allow younger generations to manage people and older generations to manage tasks.

Encourage Cooperation over Hierarchy

With the Generation Y attitudes, the old hierarchical work structure is deteriorating. Young employees are looking for leaders who will show them what to do and not tell them what to do. The U.S. Army learned this difficult lesson and changed its management style accordingly. In the past, drill sergeants gained respect through intimidation and fear. However, when Generation Y recruits came in, this strategy backfired. Up to 10 percent of recruits decided to leave the training programs, causing drill sergeants to rethink the old ways. They started to lead by example instead, acting more like mentors and doing everything alongside their recruits. The results were fantastic, with 50 percent less recruits dropping out. Generation Y does not put up with being told what to do. They will not stay long in a hierarchical system with poor treatment.

Allow Them to Work Independently and Creatively

Whereas the Baby Boomers may not complain about the work they are given, Generation Y'ers expect work that is challenging. They want to express creativity. They also do not want a leader to stand over their shoulders and monitor their progress. Generation Y'ers want the freedom to work on their own and expect trust from leaders that they will complete their given tasks. With such an efficient and talented generation of workers, leaders can be willing and excited to provide such opportunities.

Managing Generation Y requires that the principles of servant leadership be followed. Project leaders are not catering to this unique generation's every whim, but rather recognizing that Generation Y has unique needs and expectations that must be met in order to help them reach their full potential. As mentioned in Chapter 5, servant leadership is not slave leadership. Project leaders are not expected to service this generation's every desire, but rather work to accommodate their genuine needs. Some of Generation Y's expectations might not be easy to meet in regard to the organization's existing rules and regulations; however, project leaders can work to incorporate smaller policies that will help Generation Y reach their potential.

Chapter 9 Review

The latest generation to enter the workforce, Generation Y, presents a unique set of opportunities and challenges for project leaders. As the largest generation since the Baby Boomers, Generation Y is changing the face of corporate America. Gen Y currently comprises just under 25 percent of the workforce, and an additional 10 million Gen Y'ers are expected to join the ranks in the next five years.

Some of the characteristics that set apart Generation Y from previous generations include the following.

The Need for Professional and Personal Life Balance

Gen Y'ers expect their jobs to accommodate their personal lives, and are willing to find another employer if the work environment at one job doesn't satisfy these expectations. Gen Y'ers dislike rigid work schedules and micromanagement leadership styles.

Total Connectivity

Gen Y is wired. The lines between personal and professional communications are blurred. While Gen Y expects to be able to communicate with their friends and family throughout the workday, they are also willing to work during personal time (within reason).

Company Loyalty

Unlike previous generations, Gen Y'ers aren't afraid to walk away from a job with which they are unsatisfied. If a job is affecting their quality of life negatively, Gen Y'ers will look for employment that is more suitable.

If employers want to integrate Gen Y into the workplace and make the most from this generation of talented workers, they must listen to and understand their needs. Gen Y needs to know that their input is valued, and must trust that their employer has their best interests at heart. Project leaders can motivate Gen Y by providing them with opportunities to advance their skills and encouraging innovation.

This is not to say that Gen Y should be handed a free ticket to the top. This generation must earn their rewards and pay their dues just like previous generations; however, they do require understanding of their unique perspective.

10

Leadership Development

Where Do Great Leaders Come From?

Rocketed to earth from the doomed planet Krypton, Kal-EL (Clark Kent) discovered he possessed great powers far beyond those of mortal men and became known as Superman, the "Man of Steel." Most people are under the impression that great leaders also derive from such mysterious beginnings, genetically packaged with powerful leadership abilities.

Society tends to believe that great leaders—who can assess situations and make historical decisions faster than a speeding bullet, influence and persuade followers with more power than a locomotive, and perform monumental tasks as if leaping over tall buildings in a single bound—must be born with these incredible gifts.

Throughout history leaders have been chosen, appointed, and afforded position through birthright. Some of these leaders have made great contributions, while others have led their people to destruction and senseless bloodshed.

There are characteristics and personality traits that can support and serve a leader well. Some attributes might seem inherent, such as a pleasant and likable demeanor, or a driving perseverance, though when examined even these traits can be practiced and developed by someone with leadership designs.

Often people believe that merely a privileged environment and position attribute enough to attain leadership excellence. Winston Churchill derived from a military dynasty. Over a century before his birth, one of his ancestors, John Churchill, was made the first Duke of Marlborough for his performance in the War of the Spanish Succession. The presumed corollary of Winston Churchill's bloodline would be his leadership in World War II. Yet he failed twice to gain admittance to the Royal Military Academy at Sandhurst, and his own father, who was politically unsuccessful, described him as "lacking cleverness, knowledge and any capacity for settled work and having a great talent for show-off, exaggeration and make-believe." In addition, young Churchill had a severe stutter which made speaking at all a chore.

Thus, Winston Churchill found his motivation to disprove his father's stinging opinions and amend his legacy. He spent his life in study and writing to develop leadership skills. He overcame his speech impediment to become one of the greatest orators of his time.

Do you know who this resumé belongs to?

- Lost job
- Defeated in run for public office
- Started business and failed
- Elected to public office
- Sweetheart died; had a nervous breakdown
- Defeated in run for public office
- Defeated in run for U.S. Senate
- Elected to Congress
- Lost renomination
- Defeated in run for public office
- Defeated in run for vice president
- Defeated in run for U.S. Senate

The next entry on this resumé is "Elected president of the United States."

This is the resumé of President Abraham Lincoln. Abraham Lincoln overcame many failures on his path to becoming a leader. Each one seemed to further develop his resiliency. United States Presidents Bill Clinton and Barack Obama were both from humble beginnings and brought up by struggling single parents.

Great leaders are born, yet they are not born great leaders. Great leadership skills and abilities are developed. The desire to lead must exist, for becoming a great leader requires work and fortitude.

Great leaders realize they have worked to achieve their status and that their work is never done. Self-awareness and continuous self-improvement are attributes of great leaders. You must be willing to examine your strengths and weaknesses and work at leadership. By keeping your mind open to new ideas and learning, you can develop leadership skills.

Leadership Development

There is a group who successfully hones people and consistently turns out great leaders: the United States Marine Corps.

The U.S. Marine Corps has two defined pools of people for leadership consideration:

1. The pool of general recruits to develop noncommissioned officers
2. Targeted officer candidate recruitment

The qualifying process for Marine recruitment results in nine out of ten people being rejected. The general recruits must have a high school diploma, though most are not considering college. Many come from troubled homes and have a history of some alcohol or drug use. In other words, this is not a group that generally exhibits signs of great promise, yet many are transformed into effective leaders in a relatively short time period.

The group targeted for officer candidates are college graduates. While they are required to pass the same physical requirements and attend an equivalent boot camp, there is a focus on leadership training. The number of leader graduates from the Marines in such a short training time is admirable and surpasses that of all private programs.

So how are the U.S. Marines so effective at producing leaders?

Leadership Is Approached as a Team Feat Rather Than an Individual Task

A fresh Marine lieutenant is assigned to a highly experienced noncommissioned officer.

Management, tactics, planning, and daily activities are all discussed. While Marines are required to follow orders, they are encouraged to discuss concerns and question decisions. This theory of open discussion does not undermine authority, but strengthens strategy and resolve. Once perceived as a sign of weakness, a leader's receptiveness to possible assumptions of mistake from subordinates and fellow officers is a testament to the empowering truth that no one person can possess all answers.

When the expression of ideas is encouraged in a nonjudgmental environment, people are more apt to contribute. This can present unusual and even odd solutions, and those are often the very ones that succeed.

Everyone Is an Involved Team Member

This approachable stance of leadership values each Marine and strengthens that Marine's trust and awareness of his or her importance to the team. Everyone is involved in leadership development. By including everyone in leadership development, leaders are created.

Leaders Learn to Be Flexible

A Marine leader must be quick to accept not only new tactics but also new roles. Each leader and team member is trained to change roles on a team in a moment's notice. This flexibility creates an appreciation for the roles of other team members, allows them to understand how the duties of different roles affect other team members, and makes a team less susceptible to the loss or absence of a team member.

Leaders Learn the Value of Peer Behavior and Discipline

A strong team evolves with the help of peer pressure. While the behavior of employees in companies is often directed at satisfying a supervisor, Marines encourage peer motivation and discipline. Fellow Marines are dependent on one another in circumstances ranging from daily assignments to possible life-threatening situations. This dependency creates trust and certain expectations. As discipline can be dished out in peer fashion, unlike in the business world, Marines form strong commitments to comrades.

Leaders Must Plan and Implement under Extreme Deadlines

Mastering urgency is a warfare requirement for survival and success. Speed of implementation is a must, and leaders are taught how to plan with the same exigency. Action is often about timing, and hesitation can be costly in warfare.

Leaders Learn to Simplify

Marines use a "rule of three." While there may be hundreds or even thousands of possible strategies for a scenario, Marines are trained to choose the best three solutions.

These three solutions are then examined, and the best one is chosen. Marines are trained to boil things down to their most basic elements. Orders are reworded to the simplest terms.

Natural Abilities Are Developed

The Marines do not insist on conformity in every aspect as is often the impression to outsiders. Leadership development includes developing the strengths or skills that are unique to the individual. The Marines assess and utilize the natural strengths of individuals wherever possible. Some people have persuasive skills to direct people. Others have an ability to communicate confidently, team-building skills, or technical expertise. By recognizing these skills in enlisted personnel, they can assess leadership potential.

Leaders Face Physical and Mental Challenges

Marines are presented with challenges that test both physical and mental fortitude. Groups are given assignments, such as moving injured comrades out of enemy territory, over obstacles, or across ravines, that require teamwork ingenuity.

Not all of the aspects of the Marine leadership program will transfer to every business application. Business managers and executives are bound by budgets and different rules or laws, and report to multiple entities such as accounting departments, tax authorities, government agencies, customers, prospects, clients, and a board of directors. Business managers and executives often have a broader range of responsibilities that limit their time and other resources.

There are several basic principles of the Marine leadership program that can be utilized in business by leaders and teams. These Marine principles include establishing new core values based on honesty, which spawns strong trust. In order to establish Marine Corps leadership principles and reap the rewards, the foundation must be based on the truth. Lies erode trust and dilute information. When decisions are based on false pretenses, the outcomes are often undesirable.

The Marine leadership program invests a great deal in the development of recruits. Each recruit is taught to recognize his or her capabilities and problem-solving skills, and learns how to assess his or her benefits to the team. In addition, recruits are informed of the purpose and development of the team—they understand the "why" of their actions whenever possible. Development goals for a Marine are more dominant than a paycheck. Input and innovation are encouraged, and desired results are emphasized.

Leadership as a Cultural Attitude

Everyone Is a Leader

In business, successful leadership is a group attitude and set of actions and a state of mind more than any one individual. The U.S. Marine Corps has embraced this concept by giving everyone leadership training. This can benefit employees by increasing their value to the group, instilling empathy for leadership roles, and clarifying their individual importance.

Training everyone in the fundamentals of leadership enables people to do the following:

- Better manage themselves through skills for self-discipline
- Have empathy for their leaders, causing them to also be better followers or team members
- Become more responsible for their actions
- Become empowered through confidence
- Develop personal skills through self-assessment and awareness

Leadership Accountability

> In the long run, we shape our lives, and we shape ourselves. The process never ends until we die. And the choices we make are ultimately our own responsibility.
>
> **—Eleanor Roosevelt**

Leaders should be accountable for their actions and decisions, and for the morale and effectiveness of their team. Businesses should evaluate their leaders partly from their employees' level of satisfaction and performance. Leaders should know how they will be evaluated.

Great leaders will hold themselves accountable for their actions. Great leaders often walk a fine line between self-degradation and gloating. The balance of ego and constant self-scrutinizing spawned from the leadership position can be tedious, and even more so when a leader has accepted accountability and the responsibility of the impact of his or her decisions. Acceptance of the fallibility of the human condition coupled with the strength gained from the acceptance of accountability with a healthy ego can make mistakes learning experiences. These mistakes are not then shoved under the rug in denial to support a false ego, but rather seized as opportunities for learning and improving. While every error need not be held high in the public eye for scrutiny, such accountability can build stronger decision-making skills.

Accountability and honesty are two halves of a cog in the machine of responsible leadership. Dealing in the truth creates an environment that makes accountability easier. Accountability in turn makes dealing in the truth easier. When a leader is dealing in the truth, there is less likelihood that decisions are tainted by a diluted perspective.

> The best thing you can do for an employee, as soon as you know they're the bottom ten percent, is to let them know it so they can get on, adjust their lives and get themselves into the right game. That, in my view, is a kinder, gentler company than the company that winks at the truth.... Cruelty is waiting (to fire someone) until they are fifty years old with a family, a mortgage, three kids in college and a thirty year stack of performance reviews that say he's wonderful.
>
> **—Jack Welch**

Jack Welch is a great leader of the current time. Jack told his employees, "Tell me the truth and tell me early." When employees covered up mistakes or fudged over them, once Jack found out they would most likely lose their jobs. However, when employees spoke up immediately about their mistakes, Jack often discussed how they would overcome any repercussions and avoid the same error in the future.

While there are a myriad of skills that can aid great leadership, and leadership can require many skills, there are several skills of particular significance that can be developed for honing the leader within you.

Attribute Checklist

Many great leaders possess some of the same attributes. Some are stronger in certain attributes than others. Here is a list of necessary and desirable attributes for great leadership.

Accountability

Great leaders hold themselves accountable first and will admit to and take responsibility for mistakes.

Aligned Ideology

A great leader has beliefs that are appropriate to the group. You might not find a group of people who believe exactly as you do. But you can often get people to share your vision if they have a stake in the outcome and your goals are not in conflict with their beliefs.

Attention to Details

Great leaders can delegate many tasks but must have the ability to review details. In order to accomplish great tasks, you must learn to approach every task as though it were great. The devil is in the details. Most every project depends on the completion of the smallest tasks.

Life itself is dependent upon microscopic details. Neglecting to consider important details can cause catastrophic failure. Our police now solve more crimes and our court systems convict more perpetrators on the smallest fiber and DNA crime scene details. An inch and a point can win a football game. The slightest misdiagnosis can cause loss of life.

Yes, details are as important as the sum of the parts. Leaders oversee the details of their projects. Delivering superior results is not the act of taking shortcuts to avoid details.

Coaching

A great leader can help people recognize their strengths and weaknesses and develop their knowledge and skills.

Collaboration

A great leader will reach out to others and find ways to harness and funnel energy to the common good. He or she can gain cooperative effort and instill the desire in others to do the same.

Commitment to Lead

While most anyone can become a great leader, great leaders do not happen by chance. Great leaders have made the decision to be a great leader. They work at learning what is needed to be a great leader and practice to improve the skills required. Great leaders practice leadership techniques. Becoming a great leader requires a willingness to develop self-perspective.

Communication

A great leader communicates ideas, directions, and concepts with the interests of others in mind and in a way so that they will understand and act as desired. A great leader listens to people and what they are saying, digests their needs, and responds appropriately.

Confidence

A common thread that runs through all attributes of great leadership is confidence. Self-confidence supports persistence, motivation, the openness to hear opposing views, the ability to persuade others, and most aspects of leadership.

A great leader has developed self-perspective that includes recognizing strengths and shortcomings. Confidence is inspired by having an awareness of shortcomings and how to deal with them, rather than denying the existence of these weaknesses.

Conflict Management

Great leaders address conflict and see these instances as opportunities to solve issues and learn.

Courage

People do not follow leaders who are not brave. Courage is needed to believe in your visions and goals when others cast doubt. Courage is needed to convince others to follow you.

Creativity

Great leaders develop their creativity. Creativity enables them to think of more solutions to problems and more ways to encourage those around them.

Curiosity and a Sustained Desire to Learn

Great leaders have a burning curiosity to learn and discover. They see everyone as having valuable knowledge and welcome opposing views. They learn from other great leaders. Great leaders stay informed of current events and the activities of other leaders, adversaries, and allies.

Decision Making

Great leaders are practiced in decision making, and because they can weigh the consequences of variable decisions or inactivity, they are confident in their conclusions.

Delegating

Great leaders recognize the capabilities, strengths, and weaknesses of others and can effectively assign responsibilities and authority.

Dependability and Reliability

Great leaders do what they say and follow through. They can be counted on to get the job done.

Desire to Help Others

Great leaders have the desire to help others and help others lead. Most great leaders see leadership as a position of service.

Empathy

The successful leader must understand others and their problems, and express this empathy.

Flexibility

Great leaders are adaptable to change, encourage input from others, and are willing to consider changes in plans based on their suggestions and ideas.

Goal Setting and Planning

Great leaders have visions or goals. They set realistic goals based on their capabilities or the capabilities of their team. They establish step-by-step plans and assign responsibility for each duty. They define and communicate these goals, and establish timelines and plans for achievement.

Great leaders generously give others the same opportunity to exceed expectations by clearly defining what is expected of them.

Information Management

Great leaders identify, locate, assimilate, organize, and analyze relevant information.

Innovation

Great leaders are prepared to think in new ways and seek answers in uncharted territories.

Insight

A great leader can reflect on events and understand the positions of others.

Integrity

A great leader has integrity and exhibits his or her values. He or she behaves in accordance with established principles of right and wrong.

Keen Sense of Justice

A keen sense of justice will earn the respect of others and construct an environment that encourages the expression of ideas and fosters innovation.

Knowledge of Leadership Styles

Great leaders understand and use different approaches to influence and lead others.

Modesty

A great leader leads without interest in personal gain. While great leaders are quick to take accountability for mistakes (and even quicker to correct them), they do not seek out accolades for their accomplishments. They do not see themselves as authorities, but as leaders.

Motivation and Initiative

Great leaders are motivated and take the initiative at every opportunity. They are proactive. A reactive leader is often seen as a follower. Great leaders are ready and willing to accept responsibility. Great leaders are motivated about their every purpose. They expound energy and excitement in others. They create an environment in which people flourish, and they motivate others.

Optimism

An attribute of a great leader is optimism. Problems are seen as challenges of opportunity. Others are inspired by this optimism.

Organization

In order to lead effectively, great leaders know they must be organized. They will utilize technology, people, and tools to get and stay organized. They realize organization is a practiced activity and not a state.

Passion

Great leaders approach their plans, problems, and challenges with a vigor that convinces others to follow. They believe in their visions and get excited discussing them.

Patience

A great leader is patient with others and plans ahead.

Persistence

Great leaders have great determination to push on even when things don't go as planned.

Persuasiveness

A great leader has strong persuasive skills. Convincing others to follow him or her will sometimes take great effort. These skills must be developed for one-on-one situations, groups, and public speaking.

Planning/Ability

Great leaders learn quickly and know how, when, and where to apply knowledge. They work with others to develop tactics and strategies to achieve their goals.

Problem Solving Skills

Great leaders can evaluate and resolve problems quickly and effectively.

Risk Assessment Ability and Courage to Take Risks

Great leaders learn that risks are required for rewards. They develop the ability to assess risks and the risk tolerance of those they lead. Great leadership requires the ability to help others grow in risk tolerance and understand the visions or rewards.

Self-Discipline

Leaders must lead by example. The actions of a hypocrite will erode the confidence of followers and the ability to lead. A great leader establishes boundaries of behavior and abides by them.

Sense of Urgency

Great leaders have a sense of urgency about them. They have an agenda for their goals. They instill the sense of urgency in others that will be needed to reach these goals.

Sensitivity and Respect

Great leaders treat others respectfully and validate their needs. They are sensitive to the plight of others and have great regard for them as individuals. Such leaders have a pleasing personality that attracts others. They are not lazy or slovenly.

Solidarity

A great leader, while recognizing individuals, commands groups as one. Solidarity sparks the interest and earns the respect of followers.

Stamina

Great leaders must have physical, mental, and emotional stamina for the demands of leadership.

Steadfastness

A great leader acts calmly in the face of disruption or catastrophe. A great leader is resolved to see things through. He or she has an even disposition and can maintain a sense of humor to calm others under difficult circumstances.

Time Management

Scheduling and managing time are essential elements of success for projects and a great leader.

Tolerance

Great leaders are tolerant of ambiguity and adversity because they recognize that through such tolerance, useful education can occur.

Vision

A great leader looks to the future and creates plans to reach new goals.

Willingness to Step beyond the Comfort Zone

Great leaders will step outside of their comfort zones. By exercising their flexibility, they become stronger. They recognize the rewards of risk.

Working to Exceed Expectations

A great leader strives to exceed the expectations of his or her followers and to set an example to followers.

Flaws That Can Limit Your Ability to Lead

The human condition prohibits perfection, and a great leader is not without weaknesses. Certain leadership flaws can ruin leadership roles, cause damage to organizations, and nullify positive attributes; these are fatal flaws. Other leadership flaws are weaknesses that prevent that person from optimizing his or her leadership effectiveness. Most leadership fatal flaws can be overcome, or eliminated.

The majority of leadership flaws and weaknesses are not difficult to spot; however, they are often ignored or not acted upon. Poor leadership is to blame for many project and company failures. When leaders have flaws that prevent the development of loyalty or weaknesses that corrode commitment, businesses suffer. A manager or leader lashing out at employees or too weak to command is easily spotted, though many other leadership weaknesses are less apparent. A leader who does not motivate employees might skate by while the blame for poor company performance falls on outside sources such as the economy.

Recognizing leadership weaknesses and making improvements are often left to leaders themselves. The good news is that for the leader who wishes to improve leadership skills, there are many tools for doing so.

Leadership Flaws

Making Assumptions or Being Quick to Judge

Making assumptions that categorize people, or reaching conclusions without all the facts, can be fatal. For example, assuming that someone is lazy because of hearsay or one incident can lead to an aversion to assigning that person work even if he or she is the best choice for the job.

Assuming a falsehood can cause future decisions based on that information to have deadly consequences. One of the most famous flawed assumptions of war occurred at the Battle of the Little Bighorn, also known as "Custer's Last Stand."

Indian agents reported to Custer there were only 800 Indian warriors to encounter. This number came from how many Indians had left the U.S. reservation with their leader, Sitting Bull. By the time of the battle, however, estimates are that several thousand more Indians had joined Sitting Bull. Custer also assumed that the Indians would be sleeping at the time he planned to attack and they would not fight but run away. His focus was on preventing the Indians from escaping, not fighting them. Custer also seems to have assumed that he had backup, and the pack train behind him would come to his aid if needed. The title of the battle could have been "Custer's Last Assumptions."

As another example, seeking to get into the well-guarded city of Troy, Odysseus had a huge hollow wooden horse built. When the horse was completed by the artist Epeius, Odysseus climbed inside with an army of Greek warriors. Then the rest of the Greek fleet sailed away, deceiving the Trojans.

One Greek, Sinon, stayed behind. He told the Trojans that the wooden horse was safe and would bring luck to them. The Trojans assumed the horse to be a gift from the gods and pulled the wooden horse into Troy.

Later in the night, when most of Troy was asleep and others were in a drunken stupor, Sinon let the Greek warriors out from the horse, and they slaughtered the Trojans.

Assumptions, no matter how small, can at the least erode the confidence and trust of subordinates and team members. Other times, they can cause destruction. An occasional mistake that is acknowledged can be overcome. While admitting a mistake was once seen as a sign of weakness, making an incorrect assumption is magnified by the ego too prideful to take responsibility. When a leader does not admit accountability for an incorrect assumption or other mistake, this is a weakness. When a leader consistently makes incorrect assumptions, followers lose faith in the leader's abilities to make decisions.

Having Unclear Goals or Lack of Vision

The reasons why people choose or follow a leader vary. Some leaders are powerful because of their direction or vision, some leaders have followers because people have confidence in the leader's ability to follow a path or plan, and other leaders have the ability to interpret a vision. A leader without clear goals or clear communication of the plan for the team will often lose the commitment of others or create a state of confusion.

If a pilot announced over the PA to a plane full of passengers, "We will be arriving in..." and you hear a pause and rustling of papers while he asks the copilot for the flight's destination, you might lose some confidence in the pilot's abilities.

How many people would board a one-way cruise ship that advertised an unknown destination? A leader must have a clear vision and be able to communicate that vision to others. Many bills have failed to pass Congress because the bill was written poorly or the backers were unable to clarify the direction or details of the bill.

A strong statement of direction instills confidence and builds momentum. The first best way to be able to communicate a strong statement of your direction is to have one!

Fear of Delegation

Many people in positions of leadership have risen to such posts by their own efforts. Often they now have subordinates to perform the very tasks they once were charged with. However, they are often reluctant to let go of this work, as the successful performance of these very duties is partly what propelled them and their organization. This creates a leader who micromanages people and does not delegate effectively.

When a leader micromanages people, they feel distrusted. This leads such leaders to doubt themselves. Organizations cannot thrive in environments of doubt. Organizations are relationships, and healthy functional relationships are founded on trust.

When people are restricted from responsibility, this can lead to feelings of powerlessness, resentment, and loss of morale. In this case, low morale is a leadership issue, not the fault of the group.

In order for a leader to become effective, he or she must learn to delegate and assign tasks. People will become energized when they are entrusted to use their skills and knowledge to complete their responsibilities. You cannot help others learn leadership skills by denying them responsibility.

Refusing to Relinquish Power

With leadership comes power, and many leaders fail only because they either abuse their power or do not know how to use their authority effectively. Distributing power to those charged with assignments will instill confidence and respect and allow them to perform successfully. When others have authority over their tasks, they are empowered, feel they are making a contribution, and have more invested.

Lack of Human Investment

A leader can cause organizational or project failure by ignoring the needs of followers or employees, such as by doing any of the following:

- Not providing the information or tools needed to do the job
- Not taking an interest in people
- Not considering compensation adequately
- Not providing a positive environment
- Not sharing goals

People need to know they and their work are valued. Treating people fairly and showing concern for their loved ones are important. Often the greatest motivator a person has for working is his or her family.

Absence of Praise

When leaders fail to recognize great performances big and small, they miss opportunities to invest in their people, and dedication is eroded. Never miss an opportunity to praise. Leaders often issue many times more statements of criticism than praise. When this happens, focus is placed on the negative side of the motivation scale. People become more fearful and concentrate so much on avoiding criticism that their actions can become more preventive than productive. People will go above and beyond what is expected of them when they are appreciated. They will aim to do less or do the bare minimum in order to avoid increasing the chances of reprimand in an overly critical environment.

Praise might be the least expensive motivator and morale booster.

Inconsiderate Compensation

People must receive what they are worth if you expect them to value themselves accordingly. In addition, your employees must be getting their basic needs met in order to focus on their work. While employees can become fiercely committed to your cause, they look forward to being paid.

A great leader understands that people equate pay with respect. When they feel underpaid, most people will not commit and will eventually leave.

Compensation comes in many forms. Offering continuing education pay, vacation pay, health insurance, bonuses, stock options, expense accounts, vehicles, relocation expenses, and life insurance have all become popular offers from employers.

There are creative forms of compensation that employees will also appreciate. Guest speakers on site who provide business and personal information such as credit management can also be valuable perks.

Ignoring the issue of fair and motivating compensation is a leadership flaw that can prove very costly.

Absence of Incentive

Leaders must get their message of goals and direction to their people and make sure the message is understood to get everyone moving in unison. They must motive people to work together toward their goals. In business, a leader must also provide incentive.

Many people leave the ranks of employment every year for the sole reason of wanting something more, a possibility of earning in relation to their performance, and to gain a sense of satisfaction for their work.

Incentives for performing beyond quota works and should be considered. Providing incentives in the form of pay is wise, though consider that some people are also motivated by recognition.

Great leaders actively seek out modes of reward for everyone. A leader who seeks reward only for him or herself and their cronies will create resentment. Only by sharing rewards with the entire team will you perpetually prosper.

Rigidity

Rigidity is detrimental to leaders and their organizations. Leaders and their organizations can inhibit growth and stifle productivity. Red tape, bureaucracy, and rules can prevent the natural flow of energy. Great leaders must take time to review their organization's processes and systems. When there is a rule blocking productive activity, leaders must make changes. Allowing the flow of energy, ideas, and communications is healthy.

Certainly structure is required for any organization to operate. People need boundaries and direction, not restriction.

Being Overly Critical

Understanding the difference between learning from mistakes and being punished for mistakes is critical to leading. A conscientious employee will be critical enough of themselves. Criticizing respectfully in private to confirm or create awareness is important, while criticizing to punish or for any self-serving reason such as to vent frustration is damaging and not acceptable. The best way to inform someone of an error is to let them find out on their own.

Criticizing and pointing out faults will undermine your authority and reduce overall productivity. Overcriticizing is always a management or leadership mistake. Usually the criticizer has issues. If an employee is truly performing at such a level to need constant criticism, then the leader is most always to blame for the following:

- Not training the person effectively
- Not delegating the job to the right person
- Not relinquishing enough authority to the person

Showing Disrespect

A basic need of people is respect, and when they are disrespected their loyalty will diminish, their contributions will be minimal, animosity will build, and retaliation is highly possible. A fool who disrespects his or her team or employees will usually be foolish enough to believe they still respect him or her.

Being Hypocritical or Not a Team Player

A successful leader should be a role model for team members. He or she should be able to collaborate with team members. Few people want to take instruction from or follow a hypocrite. Parents who practice "Do as I say, not as I do" create rebellious and troubled children. Leaders should practice what they preach and personify the standards of the organizations they represent.

Respect is given and earned. Respect is taught by action. When people have not been taught how to show respect, they appear disrespectful. A leader has a great opportunity to earn respect by defining respect by example. Few great leaders gain respect by showing contempt.

The HMS *Hermione* was a frigate of the Royal Navy that fought in the French Revolutionary Wars. Captain Hugh Pigot was considered a ruthless disciplinarian who ordered brutal punishments for random actions. In 1797, Pigot gave orders to three sailors to reef the topsails. In an effort to hurry the men, he ordered that the last one down from the sails would be severely flogged. All three sailors were panic stricken and in a race to descend fell to their deaths. Other sailors protested Pigot's command, and he had them flogged. In a violent attack, the crew mutinied and killed Pigot and eight officers.

Leaders who rule in an environment of fear might get respect in the form of customary actions while their subordinates secretly plot against them.

Leaders must be team players in order to work well with other leaders. Many times leaders can conflict with other leaders because of egos. The word "ego" has developed a negative connotation. Leaders need a strong sense of self and confidence. When this sense of self becomes arrogant or self-serving, the ego is no longer in balance. When a person has low self-esteem or doubts, the ego is weak. A great leader needs a balanced ego.

Rigidity

A leader who resists changes or is not open to new methods, procedures, or ideas can be demoralizing to employees and a dragging anchor for an organization.

In the sixteenth century, King Philip II ruled Spain. Although Spain was a powerful and wealthy nation, the country was laden with debt. England, at the time, was a much smaller country without much power or money. King Philip assembled the world's largest naval force in history to invade the weaker foe. As he prepared for the invasion, his commander died and he ordered the Duke of Medina Sidonia to replace him. While the duke had extensive land battle experience, he had never fought on the sea. He requested that Philip replace him, but his pleas were ignored. And so, the so-called Invincible Spanish Armada set sail with over 130 ships carrying more than thirty thousand men.

One of the biggest mistakes that decimated the Spanish Armada was the flaws of the leader himself. King Philip insisted on controlling every detail of the great Armada mission from his palace in Spain. He rarely met with commanders and did not allow his experienced military leaders to craft any of the strategic plans. He refused to listen to their advice, despite the fact that he had little military training and no naval experience.

The English had created a series of signal beacons along the coasts for an efficient communications system. This allowed the English ships to position themselves behind the Armada unseen in the cloak of night.

When the Armada reached Calais, they were confronted by sixty English warships.

The English tactics and maneuvers forced the Spanish into retreat, although losses were light. However, the English blocked the return channel and the Armada was forced to sail north over Scotland. The weather deteriorated, and the Armada was battered. Less than half of the ships returned to Spain, and more than twenty thousand men were lost.

Refusal to listen or entertain innovations creates a stagnant environment where employees stop innovating. The consequential fallout can be staggering. Companies can lose profits using outdated procedures and fall behind their competitors, lose valuable employees, and lose market share as their products or services face obsolescence.

Refusal to delegate is also a form of rigidity. When a leader hoards tasks, team members might surmise that the leader

- is trying to get credit for performing tasks,
- does not have confidence in the team members, and/or
- is a poor delegator.

A leader who does not seek or consider the input or advice of outside sources is also impaired with a flaw of rigidity. While confidence in one's ideas is imperative for great leadership, the leader who perceives his or her ideas to be superior solely because of ego is weak and dangerous to the organization. A great leader realizes that no one has a monopoly on information and each individual has value, and sees people as resources. Even adversaries can offer opportunities for information.

There are not many positive conclusions that team members will draw about a rigid leader.

Poor Decision-Making Skills or Indecisiveness

While inaction can have other causes, the results of which are no less detrimental, it often derives from loss of confidence in one's decision-making ability. Poor decision-making skills are hazardous for the average person and most often fatal for the leader.

The botched handling of the response to Hurricane Katrina in Louisiana is attributed to a failure to react by leadership and emergency agencies.

Ignoring the Best Interests of the Team or Disregarding Company Goals

A self-absorbed leader only watching out for him or herself is easily recognizable and easily toppled. When self-centered or egotistical people are in a position of leadership, they neglect the development of their subordinates and ignore company goals. People become engaged when they can see what is in it for them. A leader showing interest in the progress of individuals reaps the rewards of strong commitments and improved performances.

Lack of Communication Skills

Great communication skills are one of the most important attributes for a great leader. People follow and admire leaders who can understand their needs and desires, motivate them, empathize with them, explain instructions and vision to them, and minimize their fears.

In addition to lacking the ability to listen, interpret, and express communications, there are poor communication actions and protocol flaws. These include the following:

- Not responding appropriately to communications
- Contacting people only for self-serving purposes
- Not following up
- Not being honest
- Being disrespectful
- Forgoing courtesy
- Not returning calls
- Ignoring messages
- Being negative

Responding Appropriately

The art of communication involves listening, and listening includes deciphering the information; hearing the words, the timing, and the voice inflection; interpreting body language and facial expression; judging social nuances; construing eye contact; and responding appropriately.

Listening is not just about the words that are heard. A talker's emotional state, feelings, desires, lack of feelings, social status, honesty, sincerity, and objectives can all sometimes be determined.

Some response mistakes include the following:

- Responding insensitively
- Responding without having all the facts (which can cause wrong assumptions)
- Neglecting to respond
- Responding at all
- Timeliness—responding too soon or too late
- Responding inappropriately
- Responding with conflicting facial expressions or body language

Response Confirmation

Even responding brilliantly is not valid unless the response has been interpreted by the recipient. Communication includes confirming that you have been heard and that the person you are speaking with has correctly interpreted your response.

Nonverbal Communication

Whether in a group or one-on-one, people are sending silent signals. These silent signals can often be as important as, or even more important than, what is being said. Picking up on nonverbal communications and evaluating their meanings comprise an important part of listening.

Nonverbal language is mostly universal. There are three types of body language:

- Intentional behavior
- Intentional and disguised behavior
- Unintentional behavior

Intentional Behavior

This type of behavior is the easiest to read and is intended to relay a message. A wink, shrug, and scowling expression all get the point across. This behavior can be directed at a group or individual.

Intentional and Disguised Behavior

Examples of disguised behavior can be a fake smile to disguise displeasure or a nod of approval meant only to expedite a conversation. Disguised non-verbal communication can be more difficult to determine.

Unintentional Behavior

This behavior includes body language and posture that are shown naturally or subconsciously during communication.

A Leader Can Learn to Interpret Silent Communication

As with all communication indicators, each message needs to be considered along with the signs of other indicators.

Positioning

How and where someone positions himself and his body plays an important role in determining several factors about him or her.

When a person directly faces you with shoulders squared toward you in conversation, this can translate to confidence, assertiveness or aggressiveness, attentiveness, and a high level of interest.

When a person is turned slightly, often indicated by the direction in which the feet are pointing, this can mean disinterest.

Posture

Good posture can indicate confidence and the desire to display power or indicate a defensive stance. If a person's hands and arms are at the sides and are not obstructing his or her torso, this indicates agreement with what the speaker is saying.

Head Position

How a person's head is positioned during conversation can reveal how interested the person is in the conversation. A tilted head can be expressing interest. A tilted head being supported by a hand can mean disinterest.

Hands and Arms

While hand and arm movements and positions can be signs of attitude, level of interest, or confidence, not any one is usually a foolproof indicator. For instance, folded arms are often seen as a sign of displeasure, though some people take this position when they are comfortable.

When people keep their hands and arms at their sides or in front of them on a table, this can be a sign of confidence and comfort. Sometimes people who are relaxed will lean back and place their hands behind their head.

Hands in pockets are usually a sign of anxiety, boredom, or submission.

Breathing

When a person's breathing is not noticeable, there is a good chance he or she is breathing normally and is calm. Erratic or heavy breathing not caused by a health condition can be a sign of anxiety, distress, or boredom.

Eyes

The eyes are great forecasters of feelings, confidence, honesty, and deception. Eyes that move quickly about a room or between people usually spell discomfort, distrust, and anxiety. In contrast, eyes that fixate on inanimate objects during conversation can be saying, "I am nervous." Wandering eyes can indicate inattentiveness to the conversation.

The direction of glances has been related to left- and right-brain activities. Avoiding eye contact can mean the person is insecure or in some way uncomfortable.

Eyebrows

Eyebrows are also great message senders.

- Crumpled brows usually mean anger, thought, confusion, or sadness.
- Raised brows can be because of disbelief, confusion, or surprise.

Mouth

How people position their mouth can be a presage of their emotional state:

- Lips tightly held together are a sign of anger or stress.
- Licking lips can be a sign of fear, a habit, or a message of desire.
- An open mouth is often surprise or shock.

Facial Expressions

Most facial expressions are easy to read. A smiling face can indicate excitement, happiness, pleasure, anticipation, or comfort. A frowning or scowling face indicates anger or displeasure. When someone's face is not giving any clues, you will need to observe other forms of communication.

Physical Contact

People inadvertently touch themselves when experiencing emotions. Some of these actions and their meanings are as follows:

- Touching a cheek shows consideration or attentiveness.
- Touching the nose can mean the person is dishonest about something.
- Touching the forehead is indicative of stress, confusion, exhaustion, or frustration.
- Touching clothing and hair is usually nervousness.

Actions

When a person fidgets, shifts, or performs repetitive behaviors like finger tapping, this can mean impatience or tension.

How fast a person moves can be a barometer for his or her level of comfort. Fast movement indicates anxiety, while slow and deliberate movements mean focus and confidence.

Learning how to deliver messages more effectively is essential to developing your leadership ability; however, listening and learning to read people are equally important. Focus on improving your delivery of messages, and then learn how to really hear what your team says back.

Charisma

Charisma is a special quality of leadership that captures the popular imagination and inspires allegiance and devotion. Wow! So where can you get that? Is there a prescription? Is there a magic little pill?

It is hard to imagine an elected world leader, king, premier, or president with no charisma. Unless you are a dictator, charisma is a must to become a great leader.

Although Adolf Hitler was a controversial figure responsible for horrific crimes and atrocities, his power has been assessed by some historians to have been mostly charismatic. Hitler believed his power came in his vision of a historic mission to save Germany. Early on, people recognized this power and followed him. Charisma relies on the perception of others.

People are attracted to public figures who have charisma. The attraction is often the feelings that the figure inspires within them, usually through speaking. The speaker eloquently expresses something that relates with or touches you. You feel a connection with the speaker's message. The feelings might be positive and uplifting, or anger that invokes a desire to take action. A great leader has designed the speech with a specific cause to action or impression.

Can you become a charismatic speaker? With practice, people can become charismatic speakers. There are specific qualities and methods of charismatic speaking; the proof of success will always be the reaction of the audience.

Smart

A great speaker is smart. What is smart? When it comes to speaking, a speaker who is knowledgeable about the subject matter appears smart. A speaker who is familiar with both sides of issues and can answer questions adequately or to his or her advantage also appears smart.

Energy

Charismatic speakers are alive! They can heighten the energy of a crowd with their energy. Rock bands have done this for years with songs that touch a chord with their audience, as well as dramatic energy with the aid of amplifiers, lights, and other visual effects.

A speaker with energy can seem bigger than life and capture people's attention. Energy is expressed in movement, voice inflection, volume, timing, eye contact, and speed of delivery.

Appearance

Great speakers look great. This does not mean they need to be great looking. Confidence can go a long way in making a speaker look great. Being happy in your own skin and taking the time to look your best can boost your confidence. The other aspects of a great speaker such as having subject knowledge can also increase one's visible confidence.

Learning how to move around the stage, becoming comfortable on stage, delivering a speech with confidence, timing, controlling facial expressions, and solid voice inflection will all help build "stage presence."

Confidence

The easiest way to portray confidence is to be confident. True confidence is not arrogant. True confidence is considerate.

Confidence can be developed in the following ways:

- Knowing your subject matter
- Knowing your audience
- Being familiar with your stage and surroundings
- Rehearsing your speech
- Public speaking
- Caring for your appearance
- Practice, practice, practice

Mood

Charismatic speakers can set the mood. Charismatic speakers can use humor, sad stories, inspiring stories, facts, and delivery to set the mood and lead the emotions of their audience.

Delivery

Charismatic speakers can deliver charismatic speeches. Learning how to speak fluently without stammering, with good pronunciation, and with intriguing voice inflection is imperative for developing charisma. This part of speaking is gained through repetition or practice.

Passion

You can almost never instill passion in another person about a subject for which you have or show no enthusiasm. There is great power in unharnessed enthusiasm. People become believers from being exposed to someone who fiercely believes. This type of enthusiasm or passion can be infectious and intoxicating.

Enthusiasm is developed through behavior. If you want to be seen as enthusiastic, then act enthusiastic. In the spirit of Forrest Gump, enthusiasm is as enthusiasm does.

Conviction

In order to give value to enthusiasm, the passion must have a message delivered with conviction. This is how a great speaker communicates his or her vision clearly and gets others to support it. An enthusiastic opinion can spread like wildfire. Conviction can be supported with descriptions of actions the speaker has taken or will embark on, the emotional tone of delivery, the steadfast consistency of the message, and the energy scale of the speaker.

Connection

Charismatic speakers know how to relate to and connect with the audience. They can relate their vision in terms of what it means in individual terms. A speaker with a positive demeanor that can draw a picture of his or her vision and relate that to the needs, desires, or fears of the audience can instill thoughts and feelings that will invoke action.

How can you connect with your audience? Body language, timing of speech, vocabulary, movement, eye contact, and visual aids all help a speaker develop a relationship with the audience. Yet there is also a seemingly magical connection many speakers seem to make with their audiences. This often comes from practice of all the factors that nurture and create the relationship.

Connection also comes from knowing your audience. By understanding who your audience is and what they already believe, swaying them to your point of view will be easier. Delivering a speech that begins with an opinion or belief shared by the majority of the audience to establish rapport and get them on your side is important. When you have a majority of an audience following you, the majority will usually grow and the opposition will dwindle and become submissive or silent. Few people want to confront or disagree with a crowd.

Well-Written Speech

Leaders, no matter how charismatic, cannot usually overcome a poor speech. A well-written speech that flows logically, builds anticipation, has pauses, is timed adequately, uses appropriate language for the subject matter and audience, and addresses the audience in a way that connects emotionally is a great start.

A great speech must do the following:

- Start by getting the audience's attention
- State the problem(s)
- Provide the solution(s)—vision
- Instill the action

Believability

A charismatic speaker must be believable. While connecting with the audience is a beginning, believability must be established through honesty, manners, and fairness.

Without credibility, all else is lost.

Credibility can be established by stating supporting facts before vision. The weather newscasters are notorious for practicing this one. Many weather forecasters begin with telling you what the weather did for the day before giving you their prediction for tomorrow. At the end of the day, few people are interested in what the temperature was and whether or not it rained. After all, they were there and know what the weather did.

Yet by stating the facts you know to be true first, weather forecasters establish rapport. They demonstrate they can observe and report history accurately, so therefore the idea that they can predict the future becomes more palatable.

Recognition of Weaknesses and Flaws Is the First Step to Improvement and Recovery

Beginning with a personal assessment of our flaws and weaknesses is desirable.

Feedback is essential to assessment. We all have a self-perception, and each person we encounter forms a perception of us that might not be close to our perception of ourselves. An inability to see ourselves realistically, or a flawed self-perception, can be fatal for leadership. Getting feedback about our strengths and weaknesses from as many sources (people) as possible increases the accuracy of our data. Aligning our self-perception with how others see us creates a more accurate starting point for self-improvement.

Create a list of the items you perceive to be your strengths and weaknesses. Now take this list to as many past and present coworkers, employees, supervisors, associates, clients, and friends as possible. If you are feeling brave, provide each member of your family with a copy as well. Ask each person to rate your effectiveness on a scale from one to ten, with one meaning "needing the most improvement" and ten being "perfect." Then ask them to fill in the blanks with any weaknesses they perceive about your personality and ability to lead.

Building Your Strengths

> Leadership and learning are indispensable to each other.
>
> **—John F. Kennedy**

Improving leadership skills is not all about working on weaknesses. Exercising and building your strengths are as or maybe more important.

Great leaders are not without weaknesses, though great leaders usually have more than one leadership strength. The more leadership strengths a person has, the more likely he or she can become a great leader.

What Are Leadership Strengths?

Leadership strengths are qualities necessary for effective leadership within the particular leadership environment. Leadership environments include business, political, sports, and social environments. Many qualities of leadership are transferrable between the environments, while some are more specific. For example, technical understanding and competence might not be as necessary in a recreational or sports environment.

Leadership strengths can include the following:

- Goal focus
- Desire to lead
- Inspires, motivates, and persuades
- Technical knowledge
- Problem-solving skills
- Communication

- Relationship building
- Negotiation
- Can see the big picture
- Strives to improve

Your Leadership Plan

Why do 3 percent of Harvard MBAs make ten times as much as the other 97 percent combined?

That 3 percent all had written leadership plans!

Most leaders do not have an established leadership development plan. Your leadership development plan should include how you will continue to develop your leadership strengths.

What are your leadership strengths?

What leadership strengths do you possess, and how do you work to improve them?

How can you make improvements?

Becoming a leader requires self-motivation. We need to take conscious control of our growth and learning. By making a plan and following your map, you can become instilled with confidence and energy. When you are on a path of improvement, you break the chains of doubt and mediocrity and begin an ascent to higher performance and awareness. Through recognition of our strengths and dedication to improve, we grow. Anything less is a stagnant state of discouragement.

Developing a leadership plan entails making choices. In order to make the best choices for improving your leadership strengths, you must establish what you believe in, where you are going, why you want to get there, and how you will get there.

In order to start your strength development plan, ask yourself:

- What leadership strengths do I have?
- What makes me feel strong?
- How have these strengths helped me in the past?
- How will I measure the effectiveness of my strengths?
- How could these strengths work better for me?
- How can I improve these strengths?
- What will I do this year to improve each of these strengths?
- What are my leadership goals?
- How will my strengths help me reach my goals?

Be specific as possible with your answers.

Getting feedback from those around you can help you better assess the effectiveness of your leadership strengths and obtain different perspectives. When surveying others, you might discover you have leadership strengths you have not accounted for and possibly credited some that you might not be exercising effectively.

Look at your strengths from another perspective: from what is important to you.

Chapter 10 Review

History has shown time and time again that great leaders are not born, but made. From Winston Churchill to Abe Lincoln, Barack Obama, and Bill Clinton, many of our world's greatest leaders have risen out of difficult situations to wow the world with their leadership abilities. As the U.S. Marine Corps demonstrates through their ability to turn out great leaders, leadership skills can be developed and honed.

Leadership Flaws

While there are many leadership attributes that leaders can work on developing to raise their leadership abilities, it is equally important to identify and address leadership flaws that may limit a person's ability to lead. There are some leadership flaws that can cause a leader to fail, no matter how gifted he or she may be in another area. Other leadership flaws prevent a leader from reaching his or her potential. Fortunately, leadership flaws can almost always be overcome or eliminated. Some of the detrimental leadership flaws include the following:

- Being quick to judge or make assumptions
- Lack of vision or unclear goals
- Inability to delegate
- Refusing to relinquish power
- Failing to meet the needs of the team
- Not offering adequate compensation for a team's investment
- Excessive rigidity regarding the team's processes and systems
- Excessive criticism
- Poor decision-making skills
- Disregard for company or organizational goals
- Poor communication skills

The first step to overcoming these weaknesses and flaws is identifying and addressing them. If a project leader is willing to do some soul searching to identify weaknesses and commits to overcome these flaws, he or she has already become a better leader.

Leadership Strengths and Developing a Leadership Plan

Leaders must do more than improve their weaknesses. Leaders must also develop their strengths to become even better. The process of eliminating weaknesses and improving strengths can be incorporated into a leadership plan. Establishing a leadership plan ensures that leaders remain committed to raising their game and reaching their goals.

11

From Good to Great

Leadership Motivation

Great leaders must motivate people to achieve the goals of the group.

Criticism

Great leaders must know how to criticize to motivate. Criticism should almost always be done in private. A great leader must walk the mine field of vocabulary and choose words of criticism carefully. Many words instantly put people in the defensive mode, and even constructive criticism will not register. Then the dragons of argument, resentment, and anger can rear their heads.

Beginning criticism through neutralization often forms a more positive mood and opens the subject's channels of listening. For example:

"I have some important information that, I believe if I were in your shoes, I would want to hear."

"I learned a way to do this that works well for me that I would like to show you."

"I want to learn what decision-making process you used to arrive at the action you chose."

"I wish I had had a mentor who would have shared with me what I am about to share with you."

There are many ways to present criticism. While not every situation requires such verbal gift wrapping, words can make or break a leader in nearly every situation.

A great leader needs to consider the personality, the motivation, the desired outcome of the criticism, and the future of the person he or she is about to criticize.

Each person is different and some people are unfortunately motivated by negative outcomes or consequences, while others thrive on reward and others through true self-improvement. Personally I prefer to motivate the latter.

145

Much of the criticism people receive is delivered for inappropriate reasons or with little thought to the message. Therefore, people have become accustomed to responding defensively.

Criticism has been used in the following ways:

- To punish a person
- To attempt to relieve the criticizer of accountability through blame
- To attempt to make the criticizer look better
- To control a person

Set goals for criticism. The desired outcome of criticism is a vital consideration. Imagine the perfect yet realistic outcome of your criticism. Would the person somehow change?

Consider first if the purpose of the criticism you are about to deliver is any of the following:

- To get the person to change a behavior, such as learning to think before he or she acts
- To teach the person something
- To improve the person's skills
- To be somehow self-serving

By transforming the harsh impressions of criticism into open channels of communication with opportunities for learning, your desired outcomes might become reality.

Appreciation

Great leaders must know how and when to express appreciation. While criticism should most often be done in private, praise is welcomed in public settings as well. Expressing appreciation is best done in the recipient's currency. While some people will be most satisfied with a comment of praise, others might rather receive a personal note. When possible, consideration for the person should be taken when delivering appreciation.

Appreciation is a tool of motivation, but if not demonstrated with sincerity and timing, it will have diminished value. Practicing sincere appreciation will make your praise come more naturally.

When I was young, I worked in a customer service position. This involved receiving customer complaints. My boss did not acknowledge my contribution and often told me I had a thankless position. I resented him and have often wondered how much more productivity he would have got ten out of me had he just once, instead of telling me I had a thankless job, said, "Thank you," instead.

Principles of Leadership Persuasion

Great leaders must know how to persuade. Without the skills and ability to persuade, the opportunity of leadership might be as difficult to obtain as bottled water in the desert.

Effective persuasion involves the accurate deduction of the cause of motivation. Persuasion is causing people to act. You must know what will make another person or people act.

Common Ground

Successful persuasion is built on trust. The more trust people have, the more movable or pliable they are. People seek similarity and common ground. People develop trust easier with familiarity. When people find congruent beliefs with others, they extend trust.

Look for areas of common ground and bring them to the forefront, whether these are beliefs, contacts, likes, dislikes, or ethics. Common ground reduces the barrier of influence.

Terms and Standards

By learning the standards others live by, you can talk in their terms. Listening to others will help you understand their terms and standards and relate to them from their positions. When you talk in another person's or group of people's standards, relate in their terms, and base your arguments on their standards, your chances of persuasion are greatly increased. When your goals are stated from their positions, you will have less resistance.

Consistency

Messages must be consistent if they are to be followed. Anyone who shifts positions or changes beliefs or moral standards, whether expressed verbally or through actions, will become less influential. Persuasion can take time and repetition. When the message's consistency changes, the pattern of repetition is interrupted and influence is diminished or eradicated.

Authority

From the time we are born and when we enter school, we are influenced by authority. People are persuaded by respected authority. When someone holds a position of authority or is recognized as an expert, people trust that person's opinion and are more likely to be influenced by messages from such authority.

Accountability

Becoming responsible for your actions has a positive side effect in developing trust. People trust a leader who admits mistakes and takes corrective action. Most people realize the fallibility of the human condition and that no one is perfect. When a leader assumes a position of infallibility, many people are aware that the leader is operating from an illusion that deteriorates credibility and makes that person less influential.

Reciprocity

A great leader recognizes that many civilized people have a desire to return or repay what they have received from others. Whether through feelings of obligation or kindness, whether monetary or in another form, when you help others, others will want to return the favor.

By helping others, you apply the principle of reciprocity. You show genuine interest in their business or situation, and invest in a relationship. Be the first to help others, and do not keep a tally in expectation of repayment.

Social Acceptance

Two powers of social acceptance are majority and individual (or organizational) powers of influence. People or organizations with influential power often have known and accepted belief systems and influential power. They have built trust.

Popular talk show host Oprah Winfrey's book club has put more books in the hands of adult readers since 1996 than any marketing or educational campaign. This is not just because she had the ability to choose good writers and books, but also because she had developed trust and influence through her vehicle of television.

The majority has the power to persuade simply by existence in most situations. People follow what others do. When people see a majority accepting an action or adapting an idea, more will join.

The Seven A's of Persuasion

Phil Baker's "Seven A's of Persuasion" are great principles of persuasion skills. In Phil Baker's book *Employer Secrets,* he discusses how he discovered the Seven A's of Persuasion and explains them:

- Announce
- Arouse
- Align

- Affirm
- Assure
- Assist
- Adjourn

Announce

1. Announce the direction of your persuasion. Make sure people know the subject of your persuasion.
2. Announce their needs and desires to be sure you know and understand them. Let them know what you believe. When you are selling yourself, make sure they know what they are buying in relationship to what they need.
3. Announce that you can and want to fulfill your prospective employer's needs and desires.

Arouse

Arouse within people the same emotions you get from the ideas that make you a believer.

Be enthusiastic and infectious. Instill your passion in others. This is not done in one simple step but as a principle that is practiced throughout all other principles and acts of persuasion.

Align

Align your offer with others' needs and desires. Show people the relevancy and benefits of your offer in their terms.

1. Align their interests with yours. Form an alliance with people. They should know you're in this together.
2. Align your behavior with that of your customer. Mirror their level of energy. Then to arouse them, increase your level one step at a time. It's much easier to lead people up a flight of stairs one at a time than to attempt to drag them or see how high they can jump.

Affirm

Affirm people's beliefs and objections. Validate them. Let them know when they get something right. Declare their position correct, or affirm that you understand. Congratulate them and rejoice with them when they join you, and never be a sore loser when they do not.

Assure

Assure people of your decisions. Assure other people's past and present decisions.

A statement of assurance should reinforce a concept or belief. Remembering to provide assurance is paramount to persuading, and that is why this is one of the Seven A's of Persuasion.

Consistently assure people that everything will be all right and that they are doing the right thing, with the same reasons you believe they are doing the right thing.

Assist

Assist the action needed. Be ready to help people take the steps to complete their decisions, and then assist them. If they need to sign a contract, have a pen. If they need to make a call, offer your phone. Take the initiative and help them make decisions!

Adjourn

Adjourn your efforts of persuasion at the right time. Know when to stop. One of the biggest mistakes I have seen many people make is overselling.

> Honest persuasion can only come from a believer. Persuasion from any less position is perversion (wrongly self-willed).
>
> **—Phil Baker**

Great leaders must know how to think in the interests of others.

Listening

Great leaders must develop their listening ability.

The Benefits of Great Listening

Listening helps you obtain information. People are one of the best sources of information about your business, inside your business, and outside your business.

Request feedback from employees and customers, and listen. When your listening skills are advanced, people tell you more.

You Cannot Talk When You Are Listening

I was once waiting for a salesman at a furniture store while he spoke with a couple looking to buy a living room ensemble. I listened while he told them how terrific a particular sofa was. The couple became bored with his spiel, and several times the couple asked buying questions such as "How soon can we get this delivered?" and "Do you take credit cards?" The salesman was missing opportunities by not listening and not remembering when he did listen. The couple finally said, "OK, we'll take it," but the guy kept right on selling. Finally, the couple excused themselves and left in a hurry.

As they shook their heads on the way out the door, the salesman turned to me and asked, "What can I tell you about today?"

Ask Questions

Develop your listening skills by asking people questions. Asking questions helps put you in control of conversations and opens the door for information to flow. Leaders must listen to everyone they interact with daily. Listening requires the ability to focus on the words, voice inflection, emotions, body language, eye contact, and demeanor of others. You must be aware of the limitations of language and the level of the subject's communication skills.

Asking questions

- gets the other person talking and you listening
- puts you in control
- gets you information
- helps you develop empathy
- develops trust

Ask Pertinent Questions

Ask for clarification: Ask people to elaborate. Ask them if they have anything to add. Many times, they will reiterate what they have already said in a new light.

Repeat for clarification: Repeat what someone has just said to you. For example:

"Let me see if I understand this ..."

"So what you are saying is ..."

"You're telling me ..."

Relating to Others

Great leaders relate to people by finding common ground to identify and empathize with them. They understand the socially acceptable boundaries of relationships.

They are genuinely interested in what people say, focus on the conversation, and can respond appropriately in a timely manner.

Remember What People Say

Great leaders develop their memories for names, places, events, and what people say to them. Picking up a conversation with someone where you left off at a later time, addressing people by their names, and remembering people's loved ones, situations, or concerns will impress people.

Charisma

Great leaders have charisma. Charisma helps leaders influence others. The word "charisma" is Greek, meaning "gift" or "of the divine." Charisma is personal magnetism.

While the definition of charisma is somewhat elusive in the fact that it is often associated with a supernatural or uncanny charm, people who are said to have charisma often have similar characteristics and abilities. These include energy, enthusiasm, self-control, and strong communication, persuasive, and leadership skills. While there is no exact set of instructions for obtaining charisma, these common characteristics are all developable.

Tips for Developing Charisma

Become Interested in People and Show That You Care

In today's fast-paced world, amid numerous technological distractions, few people take the time to acknowledge strangers or take an interest in them. Start by greeting those around you and asking questions. Instead of the usual "How are you?" which has been so overused with no expectation of response that this question now has no more meaning than "Hello," try asking a specific question such as the following:

"Do you enjoy your job?"

"How long have you been doing this?"

"Do you work here in the building?"

"Can you tell me the name of a great place to eat around here?"

At first these greeting questions might seem awkward, but after a few times they become natural. Showing a genuine interest in others wins them over and creates loyalty.

When people respond to you, answer them with enthusiastic interest: for example, "Terrific! I have heard of that restaurant, and though I always wanted to eat there I never knew anyone who had. Thanks to you, I am going to try it."

When you meet people, like them first. Don't wait to find out if they will like you or not. Don't wait for more information to filter so that you can decide whether or not you like them; start from a position of like. When you like others first, disliking you becomes more difficult.

Be Enthusiastic

One of the best ways to develop enthusiasm is to act enthusiastic. Smile and approach projects head on. Let people know you are full of positive energy, and they will be attracted to you like bees to flowers.

Show Optimism

Optimism when supported by a vision, facts, and a belief system will become infectious. People want to believe in something or someone bigger than life. Looking to the good side of things and refusing to be brought down by the negativity of others will move you above the crowd.

Become Open

Welcome input and advice from others without judgment. This will open channels of information flow. Criticism and judgment of others and their ideas can obstruct the flow of information and create resentment that works against the progress of the group or organization.

Passion

A substantial ingredient in charisma is passion. Your passion is a well of energy from which you can draw endlessly. This energy can excite others and influence them. Passion is energy and energy is life, and people want to believe in life. When your message is delivered with passion and your day-to-day activities and interactions with others are injected with this same passion coupled with the other aspects of persuasion, you will become magnetic.

Be passionate about your goals, your people, your clients, and your future.

Great leaders do not look down upon employees or see them as a means to an end. They become genuinely passionate about each individual's contribution and potential. This passion will ignite and inspire employees to perform. When they see that you are excited about them, they will want to live up to your expectations.

When you are passionate about your business and clients, your employees will follow. Service will improve, productivity will pick up, and profits will increase. When prospects or customers see the passion in your organization, they will buy in.

Mel Fisher was once a chicken farmer who became an undersea treasure hunter. Mel did not have the funds to pay a crew, so he recruited helpers by promising them a piece of the pie on the discovery and recovery of treasure. His passion attracted followers, and in no time he had a working crew.

People did not follow Mel solely because he might strike it rich. The work was excruciating and required long trips at sea away from home. These people could have bought lotto tickets and sat in the comfort of their home waiting to check their matching numbers. The odds were not much different. They went along with Mel because he had a burning passion and spoke with enthusiasm that excited the treasure hunter deep inside people.

Mr. Fisher did find hundreds of thousands of gold and silver coins, and valuable jewelry worth hundreds of millions of dollars from sunken Spanish ships off the coast of Florida.

Advanced Leadership Concepts

The Flexible, Nimble, and Strategic Leader

While leaders must be tough and strong, they should not be rigid. The mighty oak tree is a symbol of strength, yet can break and fall like kindling in a wind storm, while a bamboo plant can bend with the wind and survive.

> Do not believe in anything merely on the authority of your teachers and elders.
>
> **—The Buddha**

Question everything. By questioning your beliefs, what others are doing, your path, and the accepted principles of business, you keep your mind flexible and open to new ideas.

Many leaders have failed through rigid thinking and fear that questioning would weaken their stance. Just the opposite is true. When you question things, you develop strength through knowledge and your convictions become stronger.

Managing Relationships

A friend of mine, Joe, had a boss who was always nagging him: telling him how he could do things better, pointing out what he was doing wrong, and belittling him for every mistake. My friend was moonlighting part-time two evenings a week at a job that paid half the hourly rate that he was being paid by his belligerent boss. The part-time boss never pointed out a mistake to Joe. He would just tell him stories about mistakes he made and how he overcame each one.

One day, Joe's full-time job boss came to him screaming over Joe being late. He went into a tirade, telling Joe he never did anything right. Finally Joe reached his limit and blurted out, "Why the hell did you ever hire me, then? I quit." On his way out the door, Joe turned to his boss and said, "By the way, this was the Daylight Savings time change this weekend. I was early today."

Joe's part-time boss offered him a full-time job, though still at half the pay. Joe jumped at the chance and has now been a loyal employee for him for the last twelve years.

A great leader learns to manage relationships and stay in control, and recognizes every person in an organization is a part of the whole.

Challenging Others to Change

The Marines know that great leaders not only influence people to commit and follow their visions, but also push the limits of the abilities of their followers and see them as leaders in training.

When employees have been trained to take leadership roles, promoting from within the ranks is a smart and prudent way to keep people motivated and have a pool of leaders to choose from that already know your goals, operations, systems, and people. You save tremendous amounts of money that would be needed for attracting, screening, hiring, and training new leaders, and you reduce turnover. Your business is always poised for expansion, and will be propelled by your people.

Expectations of Change

If people would all do exactly what you want them to without questions, conflicts, or mistakes, you would not be needed. But leadership is about people and people management. Projects are about people. Mistakes are going to happen, and conflicts are going to arise. While you can challenge people to change and help them improve, do not set the bar too high. Accept that the job of a leader includes the tasks of delegation and of people and relationships.

When mistakes and conflicts happen, these should be your opportunities to excel. Learn new skills to better manage people, and look forward to mistakes and conflicts to practice and hone your leadership abilities. Every financial crisis, project delay, and emotional client is the chance for you to improve. Every such experience will only make you more able to handle the next one. Soon your leadership skills and style will set an example the people around you will admire.

Some of history's greatest leaders would be virtually unknown today without times of trouble. George Washington, Abraham Lincoln, Martin Luther King Jr., and John F. Kennedy all led during times of war and unrest.

A harmonious workplace and team is the goal, though you should never shy away from those times that test your leadership ability.

Perception

Think back about people you know well now, and remember when you first met them and your first impression of them. How different is your opinion of these people now?

Write down three misconceptions or judgments you had about each person. What caused your thinking? What misperceptions do you think your team members have made about you?

Create the perception that others will have about you. By exhibiting confidence, you will appear professional. Dress, act, and speak like a leader, and people will see you as a leader. A leader is as a leader does.

Place the same confidence you have in yourself in others. Let them know you believe in them. When someone believes in us, our confidence increases. Do not allow your perception of people to underestimate their intelligence or capabilities. Give them a reputation to live up to. Form the belief that people are smart and good.

Learn to Lead Selflessly

By leading selflessly, you will lead selfless people. A leader is an example of behavior to be imitated. When people like and respect their leader, they tend to imitate the leader. Even clients will be affected by a great leader.

Have a Purpose

People want a purpose. They thrive on working toward a defined goal. Define your company goal, your project goals, and goals for individuals.

Getting Others to Work Together

No matter how developed your leadership skills become, you will not be able to force people to like one another. However, you should be able to get them to work together for the greater good. By showing them the possibilities of confidence and rewards of working together and what's in it for them, you can get people to go a long way.

While people might complain, and while chronic complaining is another issue, a great leader can recognize moments of discontent as opportunities to gain valuable information regarding people's frustration.

Build Your Team

Prepare your team and make your team prepared. Surround yourself with the best people you can afford, and train them as well as possible.

Integrity

While integrity is a basic characteristic for a great leader, practicing integrity can involve wrestling with complex decisions almost daily. Strong convictions can make many decisions easy, though leaders are often faced with more perplexing challenges.

Integrity means consistently abiding by your standards in thought and action. Adhering to values, goals, and ideals in every action for yourself and as a role model is critical to great leadership.

Courage

Great leadership requires courage. This is the courage to make difficult decisions, face conflict, stay the course, and blaze a trail for others to follow. Courage comes from confidence, and confidence comes from knowledge. Success breeds continued confidence.

When others around you see you exhibiting confidence in the face of adversity, they will fall in behind you. You are the example for your followers. Boldness is another attribute of immeasurable value.

Your courage should be exhibited in all tasks, large or small. Complaining erodes the impression of courage, and an attitude of vitality to get the job done will strengthen your example.

Great Strength in Humility

Great leaders distribute the credit for a job well done and remain modest about their contribution. They know this practice creates loyalty and dedication that will carry them forward. When employees receive credit for their contributions, they perceive their value.

You can share in the pride of your team's successes but pass the glory whenever you can. This can be difficult to do at first, but when you see the results you'll find the long-term rewards are often far greater than if you hogged full credit for every win.

The Business Is People

People are of higher value in the workplace today than ever before. The average worker requires more skills, and the cost of turnover is high. A great leader knows that his or her vision and business depend on attracting, recruiting, managing, and keeping good people.

Take a genuine interest in the people on your team and their likes and dislikes, personal lives, and beliefs, and this will spawn deep loyalty and give them a higher sense of worth and job satisfaction. In addition, knowing more about your people will give you a deeper understanding of them and more leverage to motivate them when needed.

Stay aware of your people's needs. People have changes in their personal lives and emergencies. Learn to exercise consideration, and be flexible in their time of need. Kindness is an invaluable attribute of great leadership.

Learn to Enjoy People

> All being said, I like only those people who are useful to me and only so long as they are useful.
>
> **—Napoleon**

If you are in the business of using people, you will eventually fail. Enjoying people might seem like a basic leadership skill, but can be an acquired taste for many leaders and entrepreneurs who are used to going it alone. The adage that it's lonely at the top might symbolize the boundary a leader maintains between a personal and business or leadership relationship, yet there is no reason for a leader to be lonely. A great leader must lead yet can enjoy the company of his or her people. Leaders who do not enjoy people or work well with others are not doing a good job.

Investigate how to involve people with your goals. A great leader learns how to help people reach their goals and discovers their strengths and weaknesses to help them contribute to company goals. By listening to your employees and their goals, you will find more attentiveness from them for your goals. Learn their strengths and weaknesses.

Involve your team when setting goals. Their input can offer invaluable information that will allow you to reform your goals. Accept all input without comment or criticism to create an open flow of energy, and you might be surprised how helpful their information is.

Focus on the people, and the projects will move smoother. When the project takes on a life of its own, the activity can dominate the attention and reduce the significance of the contributors. Do not less this happen. Make sure people know they are more important than projects.

By focusing on people, you will free them of many worries and conflicts, reducing miscommunications and mistakes and empowering them to operate at a higher capacity for you.

Sharing Information

> If people are informed they will do the right thing. It's when they are not informed they become hostages to prejudice.
>
> **—Charlayne Hunter-Gault**

In order to empower your employees and enable them to make better decisions, keep them informed. While there might be some confidential information, hoarding too much information forces employees to make assumptions. Sharing information validates the worth of your people and their importance to your team.

Sharing information makes people feel secure. Hiding information generates theories of conspiracy and pits people against each another. Information can become a weapon to wield rather than a tool to produce. Suddenly you have a team wasting valuable time planning and making manipulative maneuvers just to get in the know.

When people have a lack of information, sometimes they fill in the blanks. They might assume an incorrect fact that causes damage to others or dilutes important information.

Employees might become reluctant to seek the information they need or confront you. Employees should be able to ask any question on their minds. They should not be thought of as nosy or up to no good.

Feed Your Culture

A company and its employees develop their own culture. Management and leaders should not fight to extinguish company culture but embrace and feed it. A strong company culture can encourage loyalty, improve working relationships, and make people feel they belong.

Company culture is malleable. It can be formed and shaped by informal after-work gatherings, Christmas parties, and outside work activities such as sports teams or events.

People who play well together work well together. Prohibiting alcohol at company functions is advisable. Alcohol can distort people, influence their actions, and destroy relationships and teams in an instant.

Chapter 11 Review

Motivation

One of the most important things a leader must do to move from good to great is to learn to motivate and inspire his or her followers. Motivation can come in many forms, including criticism and appreciation. In some cases, constructive criticism is all that is required to encourage a team member to raise his or her game. For other team members, an expression of appreciation might have a more powerful impact. There is a time and a place for both criticism and motivation, and leaders must use these tools appropriately for maximum effect. When a leader understands the best way to motivate followers, he or she will be able to lead them to achieve bigger and better things.

Persuasion

Not every situation requires a leader to motivate his or her followers—in some cases, motivation is impossible. In these circumstances, leaders must rely on their powers of persuasion to get things done. Leaders must be skilled in the art of persuasion so that they can cause those around them to act.

Listening Skills

Great leaders also possess incredible listening abilities. They know what questions to ask, they are able to relate to others, and they remember what other people say. The ability to ask pertinent questions and remember the way people respond benefits leaders tremendously.

Charisma and Passion

Charisma is personal magnetism, and it enables leaders to influence, persuade, and lead others. One of the biggest elements of charisma is passion. Passion creates energy that can be used to inspire and excite teams. When great leaders are truly passionate about each team member's potential and work to communicate that passion through their charisma, people will follow.

It's All about the People

At the end of the day, great leaders are defined by the people who follow them. Have those people improved as a result of following? Great leaders have the courage to inspire change and incite conflict. They have the ability to unite their teams behind a common vision, and the strength to keep pushing the team forward when times are tough.

Great leaders possess integrity, humility, and flexibility. They understand that they are defined by their teams and do everything in their power to work through their teams and help them to succeed.

Section 3

Getting Things Done as a Project Leader

12

Project Leader Organization

Great leaders are almost always great simplifiers.

—**General Colin Powell**

As you know, high-performing teams must be built upon a strong team triangle in order to achieve the greatest heights. The team triangle is composed of three equally important sides: the individual, the task, and the team. Without a strong base on any side of the triangle, the team's ability to successfully complete a given task or project is severely hindered; therefore, project leaders must work to develop each side of the team triangle constantly and in equal measure.

We've already defined the needs and expectations for each side of the team triangle (and will summarize them again for you here), but project leaders need organizational systems and processes to help meet these needs and move their teams toward the finish line. Leaders need practical systems they can implement to control and direct ever-changing projects, while continually developing all sides of the team triangle for improved efficiency in the future.

In project teams, the "getting things done" philosophy becomes vital. Team leaders and members are constantly overwhelmed by the amount of work and tasks in need of consideration and completion. Having a common structure that enables everyone to filter information, ideas, and tasks into the appropriate category is crucial to staying on target within a specific project and maintaining the health of the team. Getting things done revolves around developing lists and processes that will help you deal with vast amounts of input quickly and efficiently.

The first section of the book focused on the need for and benefits of a project leadership mentality. The second part discussed the styles, attributes, abilities, and developmental processes of an extraordinary project leader. Now you will learn specific systems and mind-sets you can implement today to improve the productivity of your team. In these final chapters, you will discover the blueprint for getting things done at each level of the team triangle: the individual, the team, and the task.

Replication on the Individual Level

The organizational systems presented in this chapter are geared toward the project leader, but should be taught to and embraced by every individual member of the project team. The project leader is, after all, an individual unit of the larger team.

Extraordinary leaders benefit from delegating results, not processes. They do not tell their followers exactly how to complete something, but instead describe the desired outcome. If your team members understand what results you expect from the implementation of an organization process, they should be encouraged to modify the organizational methods presented here to best suit their needs. In other words, do not force your team to follow your method of organization exactly. If they are delivering the desired results (i.e., the ability to process information quickly and get things done), some variation on the details is perfectly acceptable.

Individual Needs in the Team Triangle

The individuals who form a team determine the success of the project. In order to perform at their highest level of efficiency, individuals on the team, including the project leader, must exhibit the following traits:

- Commitment to the project leader, the team, and the task.
- A high level of energy. Increased energy will enable individual team members to improve their performance in terms of speed and quantity.
- Orientation toward results, which will enable the team to focus on completing the task in accordance with the predefined success criteria.
- Autonomy in the approach to the project. If individuals are committed, energized, and focused on achieving the desired results, every team member should be able to express creativity in how they execute the agreed-upon plan.

These basic needs must be met on the individual side of the team triangle in order for a team to achieve the highest level of performance. A good team leader will not only possess these characteristics, but also understand how to bring out these traits in their team members. In the pages that follow, you will learn how to get things done, as a leader and as an individual unit of your project team.

Getting in the Zone

To some extent, being a great leader requires that you maintain a sense of calm and remain relaxed in the face of incredible challenges. But with the constant influx of information and our ever-evolving to-do lists, how can anyone be expected to keep their cool? How can a leader maintain composure and control while thinking and working at the highest level of productivity?

The majority of people are unable to react to things at the appropriate level. Rather than responding to e-mails and phone calls at a specific time set aside for answering correspondence, they allow these interruptions to intrude upon their day like a constant parade of distractions and respond to each stimulus as it occurs.

Any nonemergency activity, idea, or input that forces you to immediately respond or react siphons your control and prevents you from accomplishing what you set out to achieve in that moment. Certainly there will be unexpected occurrences that demand your attention. In fact, much of the need for excellent organizational systems stems from the fact that leaders must be free to respond to urgent, unexpected, or otherwise critical situations at a moment's notice. Leaders absolutely must be prepared for and able to cope with the unexpected. If a leader is excessively overwhelmed by the minutiae—e-mails, phone calls, meetings, and other regular responsibilities—he or she cannot be ready to tackle the tasks that couldn't be anticipated coming down the line. Leaders should have a system in place that automates their daily responsibilities in a consistent, efficient manner. That way, when the unexpected occurs, they can trust that everything else will be under control while they respond to the emergency or unexpected task.

Allowing an unexpected crisis or emergency to break your focus is one thing; however, if e-mails, phone calls, or angry clients send you into a tailspin, you will only lose your concentration and enter into a reactive state that limits your ability to get things done. Even things to which you do not physically react can blow your focus. If nagging guilt about missing your daughter's soccer game, worries over the delinquent credit card bill, or concerns about your mother's ability to live independently constantly loom in your mind, you will not be able to concentrate on the task at hand, thus reducing your efficiency.

Have you ever been so focused on a task or activity that you lost all sense of time and place? You most likely felt as if you had things completely under control and, at least for a time, felt any stress and anxiety about the project slip away. At the highest levels of concentration, even hunger pangs, drowsiness, and other distractions can be ignored in favor of continuing to plow forward through your work. During these periods of intense productivity, which artists and other creative minds refer to as being "in the flow," we reach our maximum productivity levels.

If project leaders could consciously force themselves into this mind-set at the start of each workday, they could accomplish far more work in less time. Unfortunately, few people have control over when they enter the flow—and once the flow is broken, it's difficult to rediscover that peace of mind so conducive to getting things done.

Most of the time, people operate in a place that's the opposite of flow—a place where they experience anxiety, boredom, loss of control, and constant distractions. This is not a mind-set conducive to optimum productivity and efficiency, and continual distractions will make even the best organizational systems worthless.

Great leaders can learn to control when they get in the flow—if only they can learn to control the flow of information around them.

Why Most Attempts at Organization Fail

To control your ability to get in the flow, you must identify a way to eliminate the distractions that force you to react or disrupt your train of thought. You can only achieve the highest levels of productivity by clearing your mind and learning to laser in on the task you are currently undertaking. Rather than focusing solely on the task at hand, most people's minds are spinning in circles, constantly considering the infinite list of to-dos and what-ifs that paralyze their productivity. It's impossible to think on a higher plane when your mind is bogged down by to-dos. Sadly, most items on these neverending checklists are merely the unfinished business that clutters people's minds.

The thoughts I refer to as "unfinished business" linger in your brain because you haven't filed the information where it belongs. Most of the unfinished business that we think about is actually irresolvable because we haven't figured out what the optimal solution is, much less the next step we need to take to achieve that outcome.

There is not an organizational theory in the world that will work if your brain is cluttered with unfinished business. Any efforts at instilling order into your routine through a comprehensive organization system will fail simply because you can't put these thoughts onto a functional to-do list. These nagging items in your mind will continue to cause stress and disorder because the information hasn't been categorized or defined. Your brain doesn't know what to do with it!

In short, unfinished business stays unfinished and clutters our minds because of the following reasons:

- We don't know what we really want to happen (the ideal solution).
- We don't know what to do next to get toward the ideal solution (a clear action step).
- We have not made note of what we need to do next in a place where our brain is positive it won't forget (a fail-proof reminder).

The majority of unfinished business cluttering our brains are really tasks for which there is no next defined action. They are not actual to-do lists of items that can be checked off, but rather little sticky notes that remind us of daunting tasks in need of completion. For example, a constant worry about your mother's ability to continue living on her own does not really have a defined next step or a neat conclusion; it's just a nagging reminder of a looming problem. "How is Mom really doing?" does not belong on a to-do list because you haven't yet decided what you want to do or how to do it.

A great deal of the information we encounter on any given day winds up in our brain as unfinished business. When we've been exposed to a piece of information, an idea, or a problem, a small part of our brain remains committed to tying up the loose ends, solving the problem, or relaying the information to the appropriate person. Sometimes, our subconscious files these thoughts away for future reference, but you can bet that they will reappear on your mind when you least expect it. Once unfinished business gets in your head, it won't escape until the matter has been resolved.

Until you have decided what you want to happen and what you need to do to get there, unfinished business will keep weighing on your mind and nagging you at unfortunate times. Furthermore, your brain will continuously remind you about this unfinished business until it can rest assured that you will not forget. Have you ever had an important meeting in the morning for which you were nervous about waking up to attend? If your mind isn't confident that your reminder—the alarm—will actually wake you up, your brain will literally rouse you from sleep every few hours to make sure you do not miss the event. The brain works the same way with unfinished business. If your subconscious does not trust that your reminders to yourself are satisfactory, it will keep bringing the issue to the forefront, interrupting whatever activity you are currently performing, as a backup.

Taking Care of (Unfinished) Business

Most people attempting to organize their lives are simply reordering useless lists of unfinished business. Until this business is settled, they will be unable to focus on instilling order to the rest of their lives and organizing the tasks that matter.

To really clear your mind, you must first identify all of the unfinished business looping through your mind. Then, you must determine what would be an optimal outcome for each situation. For instance, would you be more comfortable with your mother's living situation if she had home health nurses making daily visits to check in and help her around the house? Once you've determined what you want to happen, you must determine the next step to achieve that desired outcome. If home health nurses would make you feel more at ease about your mom's health and happiness, the actual to-do item for your list might be to contact her insurance carrier to verify she has home health coverage or begin interviewing home health care providers.

The last step in eliminating unfinished business from your mind is to write down the actual to-dos for each issue in a place your brain trusts it won't forget. Just thinking about what you want to do next is not enough. Without reminders in a trusted place to count on, your brain will feel the need to keep bringing the problem to your attention.

In the next sections, you'll learn a system to help you deal with unfinished business and the other information and input you receive throughout the day in an efficient and stress-free manner.

Developing a Workflow Management System

Through extensive leadership development, perseverance, and skill, you can become a better, more efficient leader. You can learn to delegate tasks, resolve unfinished business, and work smarter to accomplish more in less time. You can create an organizational system that enables you to identify, categorize, and complete any task, idea, or responsibility that comes your way. In order to do this, however, you must have a system for handling everything that comes your way.

The majority of the workforce in this world isn't struggling with a shortage of time, even if many people believe getting that an extra hour in a day would solve all of their problems. The issue that most people are confronting is an inability to identify the next action they must take to move a project or task forward. This is where an effective workflow management system comes in. Every leader in the world should have a system in place for filtering information, making decisions, and identifying the next action necessary to move forward.

To get organized, you must have a system for processing all of the input that comes your way. This system must enable you to quickly filter information into the appropriate category, make decisions as to the desired outcome, and define the actions that must be taken to move forward. This system must be able to process every piece of input you encounter in your personal and professional life, whether you are managing two projects or ten.

The Workflow Management Model

Every piece of input or information you come across must be filtered through every phase of this system in order to prevent it from becoming unfinished business that preoccupies your mind and prevents you from focusing on your work.

The phases of the Workflow Management Model are as follows:

- Assemble
- Analyze
- Arrange
- Assess
- Act

Your workflow system also must have a mechanism for ensuring your brain is not overworked trying to remember all of the information, decisions, and actions you encounter in a given day. In other words, you must have a system for capturing the results of your workflow system so that you can actually get things done. Before we tackle the organizational method you use to record the to-dos that result from the Workflow Management Model, let's explore each phase of the system in greater detail.

Assemble

It's important to understand that you cannot just "organize" unfinished business or incoming input, because you do not yet know how the information fits in the bigger picture. This information must first be stored so that you can process and organize the necessary action steps for each item. In the first phase of the Workflow Management Model, you must assemble all of the information and input you encounter in a reliable place for future review. In order to make a decision regarding the desired outcome or specify the next action step for any single task, you must first completely understand the scope of all the issues, information, and input you are facing.

The method you use to assemble the input must be able to capture everything that demands your attention. You are already assembling some input. Messages in your voicemail, e-mail, and mailbox are currently piling up. You probably have junk drawers filled with little gadgets in need of batteries or repair. Your computer monitor might be covered in sticky notes—or you could have drawers stuffed with legal pads, notebooks, and other random pieces of paper that you've accumulated over time. Now is the time to start assembling all of this stuff—papers, messages, odds, ends, and all—into one place. You must gather everything that you need "to do" so you can figure out what actually needs to be done.

Choose Your Collection Containers

Every single thing that requires your attention must be assembled into "collection containers" so that you can administer, arrange, appraise, and act on them later. These items can be stored in nearly any kind of collection container, as long as the collection container is in a safe, reliable place outside of your head. Your collection container might be any of the following:

- An in-basket
- An e-mail account
- A notebook
- A digital recorder
- A computer spreadsheet or word-processing document
- Any combination of the above items

You can use any one or more of these items as collection containers, so long as you consistently use the container(s) to store your stuff until you are ready to analyze it. Choose one or more of the above types of collection containers to start gathering everything you must process. Make at least one of the containers extremely easy to access so that you can send or place reminders to yourself from anywhere, anytime. For instance, you might use an e-mail account as one collection container and simply send yourself messages as necessary. You could also carry a folder or notebook for recording information you encounter throughout the day.

Your collection container must meet five criteria to succeed:

- Every single piece of input, information, and unfinished business you need to consider or address must be stored in a collection container.
- You must use as few collection containers as possible, but enough to meet all of your needs.
- Your collection containers must be a regular part of your life where you consistently file information for processing and review.
- Your collection containers should always be accessible.
- The collection containers must be cleared on a regular basis.

The final criterion is of special significance. If you do not clear your collection containers periodically by moving to the Analysis phase of the Workflow Management Model, the system will fail, and your collection containers will become just another place to store your unfinished business. You do not have to empty the collection container every time you place something new in it or even every day, but do not allow the containers to overflow or they will become worthless.

Analyze

The next step of the Workflow Management Model is to analyze the information stored in your collection systems. Getting everything you need to address out of your head and into a collection container is a great start, but information sitting in a storage system is no closer to completion than it was while stored in your head.

You should frequently clear out your collection containers for analysis. During this processing phase, you can clarify the items demanding your attention and determine the necessary action steps to move forward. During the analysis process, you determine what, if anything, needs to be done with this particular piece of information. Later, in the Arrange phase, you'll determine how and when to get it done.

Easy Item Analysis

To analyze the items in your collection containers, follow this simple four-part series of questions and prompts.

Step 1: Define the item.
- What is it?
- How does the item affect you?
- Why is the item in your collection basket?

Step 2: Determine whether you can actually do anything with the item.
- Is there an action that can be taken?
  - If yes, move to Step 3.
  - If no, the item should be filed in one of four places:
    - In the trash
    - In a future action file
    - In a reference file
    - In the appropriate project information file

Step 3: Define the next action to move the item forward.
- What do you need to do next to move this item out of your collection container and onto your to-do list?

Step 4: Determine how long it will take to complete the action item determined in Step 3.
- If the task will take five minutes or less, do it immediately.
- If the task will take more than five minutes:
  - Delegate the item to someone else.
  - Write a note on your "pending" list (find more details on the pending list later in this chapter).
  - Schedule the task.
- If the action item is critical (but requires longer than five minutes), add the item to your "urgent" to-do list.

- Otherwise, add the task to your calendar for a future date.
  - If the action relates to an ongoing project, record the action item on the "projects list" specifically for that project.

This four-step analysis should be completed for every single item in your collection baskets with a few very important rules:

1. Nothing can be put back into the basket! The collection container must be completely emptied. This is not to say you have to immediately do everything that you've just analyzed, but you simply must define each item and determine what you're going to do with it from this point forward.
2. You must analyze the items in the basket in the order they appear. You cannot pick and choose the most interesting, fun, or easy items to process first. Create a pile of the items and tackle whatever lands on top first.
3. You must analyze one item at a time. Do not let your attention shift to the next item in the basket simply because you are uncertain about what to do with the first item. If you want to get the unfinished business out of your life for good, you have to make decisions about the items in your in-basket! Don't allow yourself to cheat on Rule 2 by focusing on multiple items at once.

To review, at the conclusion of the Analyze phase, you should have done the following:

- Eliminated every item in each of your collection baskets
- Recycled or thrown away anything that you do not need to act upon or keep for future reference
- Finished any items that take less than five minutes to complete
- Delegated or scheduled nonurgent tasks
- Sorted (or immediately filed) action items and nonaction information for processing in the Arrange phase

In the Arrange phase of the process, you will learn how to handle the action items you have deferred, delegated, or scheduled for future completion.

Arrange

All of the items in your collection system should ultimately end up in one of eight categories:

- Items for which there is no action
  1. Trash
  2. Future action files
  3. Project information files
  4. Reference file
- Action items
  5. Urgent to-do list
  6. Project lists
  7. Pending item list
  8. Calendar

Let's discuss these categories in greater detail.

Non–Action Item Categories

Trash

If an item in your collection basket has no required action, has no value as a reference material, or does not belong in your future action file, trash it, delete it, or recycle it. For example:

- A flyer for an event that has passed and is of no value for future reference. Because you have no use for the information, you throw the flyer straight into the recycling bin.
- A mass e-mail from a former colleague soliciting sponsors for an upcoming 5K run. Because you have a personal rule against making donations to members of your team—because you would then feel obligated to give to everyone who asked—you delete the message.

Future Action File

While placing a file with the word "action," a future action file does not have to reflect only actions, but can also represent ideas for the future. In other words, these are items that you would like to either do or deal with in the future so you put them in the file so you do not lose track of them. For example, if your collection basket contained information about a conference that you cannot attend this year but would be interested in participating in next year, you could file the details in your future action file. Your future action file might also include lists of books to read, movies to rent, dream vacations, classes to take, or ideas for Christmas presents for the family. For example:

- A magazine clipping with a quote you would like to use on your mother's birthday card. The clip belongs in the future action file since you have determined a specific use for the information at a later date. You also might add a note on the calendar entry for your mother's birthday to remind you that you wanted to use the quote.
- A sales brochure about new software you can use on a project that begins in six months. The brochure belongs in a future action file because you know you will need the information at a specific date in the future. Again, you could make a calendar note to remind you about the information near the date you expect the project to commence.

Project Information Files

Project information files work hand-in-hand with project lists (see below). Some items in your collection basket may not require an action, but the information is needed for future reference on a specific project. Any supporting documents, information, details, plans, or notes relating to a specific project should be filed in the appropriate project information file. For example:

- A newspaper article pertaining to a current project. The article contains useful information but requires no immediate action; therefore, it should be filed in the appropriate project information file for future reference.
- A phone message taken by a coworker regarding new details about a project. The client sent you the information as an FYI but does not expect a return call or follow-up conversation, so you file the information in the relevant project information file.

Reference File

Anything that does not pertain to a future action or current project that you need to keep belongs in the reference file. The reference file is a place to store ideas, information, or other data that do not directly correlate to existing projects or plans. This file should be broken down into multiple categories (e.g., personal, travel, education, trivia, and interesting articles). For example:

- A former colleague sends an e-mail with information about his new gaming website. Because the website has nothing to do with an existing project or plan, you file it in your entertainment category of your reference file so that you can check out the site later.
- A database of contact information for sales reps from whom you frequently purchase software and office supplies. The database belongs in your reference file because you don't need the information immediately, but will need to access it in the future.

NOTE: The project information and reference files are not mere "hold for future review" files where you can shove anything that doesn't seem to fit in another category. The project information and reference files are the tools that will enable you to complete the projects and tasks you are planning and preparing for.

Action Item Categories

Urgent To-Do List

The items on this list should represent critical actions you must take to move items in your collection basket closer to completion. This list should be kept as short as possible—whenever possible, action items should be delegated or calendared for future completion. Your to-do list might also be broken down into categories, for example: calls, meetings, errands, information to review, and the like. This will enable you to complete similar items in groups, which increases your efficiency. For example:

- Your boss forwards you an angry letter he received from a client and asks you to calm down the client about the status of his project immediately. Because the item is of an urgent nature but will take more than five minutes to complete, you add a reminder to your urgent to-do list under the heading "Calls."
- Your daughter's school calls and says that she must be picked up within the hour because the nurse believes she has the chicken pox. Since you have to pick your daughter up immediately, the task belongs on an urgent to-do list. You may also have to add an entry for "Find someone to watch Maggie from 12 to 5" to your list.

Project Lists

Any action that pertains to an ongoing project should be recorded on a project list, in addition to being placed on the urgent to-do list or calendar when necessary. Some tasks require more than one action, but are not actually related to a specific "project." For example, you might not consider hiring a new employee to be a project, but several unique actions must be completed in the hiring process. For this reason, it is helpful to create a project list for any task that will require related actions in the future. The project list will enable you to see what action items need to be done now, as well as what's coming down the pipeline in the future. The project list differs from the project information file in that items on the project list require action (but not immediate action), while items in the project information file are simply there for future reference and review. For example:

- A colleague calls to tell you that he ran into one of your current clients at a ballgame the previous evening and the client requested that you call him to review the project's budget at your convenience. The request is not urgent and pertains to an existing project, so you write a reminder to yourself on the appropriate project list.

- You are invited to speak at a workshop as a local expert as part of a professional speaker series in six months. The coordinator asks you to send a short biography in two months and a copy of your presentation materials in three months. Because the workshop will require multiple similar actions on your behalf, you create a project list specifically for the workshop and add the items to your list. You also add reminders to your calendar near the deadline for each item as an additional reminder.

Pending Item List

When you delegate a task to someone else, file a piece of information that relates to a future action, or have another piece of critical information for which there is no immediate task, the item should be listed on your pending item list. This will serve as a reminder of any unfinished business that you've delegated or deferred. Your pending item list should reflect anything that you are waiting on or for in order to complete a future task. For example:

- You ask your assistant to pull all of the opposition research about a specific competitor so that you can review the information before an important meeting. You need the information to complete the action item on your calendar ("Review competitor research"), so you make a note of your request on the pending item list in order to follow up with your assistant later.

- You submit a press release to the higher-ups for review, knowing that your boss has a tendency to lose or misplace documents. You can't distribute the release or make follow-up calls until you have your boss's revisions, so you add a reminder to your pending item list to retrieve the press release from your boss in a timely manner.

Calendar

The calendar is the most critical of the eight categories into which items are sorted during the Arrange phase. The calendar should reflect three specific things:

- Meetings, appointments, calls, action items, or other obligations that must be completed or attended at a specific time.

- Meetings, appointments, calls, action items, or other obligations that need to be completed or attended on a specific day.

- Information that you need to know pertaining to any of the above meetings, appointments, calls, action items, or other obligations. For example, you might include directions, maps, contact information, background information, or other details about an event or activity.

For example:

- You receive a reminder of your yearly checkup from your dentist. Because the event already has a specific date and time, you add the appointment to your calendar immediately.
- Your cousin sends you a "save the date" for her daughter's wedding, along with information about the upcoming bridal shower your aunt is hosting. You have the event details, so you schedule the item on your calendar. Because the shower invite contained information about the bride's registry, you include the stores she has registered at on the calendar entry.

NOTE: Because the calendar is so central to the success of the Workflow Management Model, additional sections have been dedicated to the topic. In Chapter 13, you will learn to calendar with maximum efficiency.

Assess

Phase 4 of the Workflow Management Model is a time for assessing and reviewing the results of the Analysis phase. At this point, your to-do list should be updated and reflect your high-priority activities, your calendar should detail upcoming obligations, and your project lists should include information about any actions that need to be completed to move forward.

The Assess phase is perhaps the most critical stage in the Workflow Management Process, because it enables you to get a big-picture view of everything happening in your life. The Assess stage presents an opportunity to examine your to-dos and determine where you stand overall on your various projects.

Conduct Daily Assessments for Optimum Efficiency

To assess your daily activities and schedule your time properly, you must examine each of the categories you created during the Analysis and Arrange phases of the Workflow Management Model. Each day should begin with a review of your calendar, followed by an examination of your urgent to-do list. Then conduct a quick scan of your project lists to ensure that everything is on track. The remaining categories, which include the pending item list, project information files, and future action files, can be reviewed during your weekly workflow assessment.

Weekly Workflow Assessments

In addition to your daily assessments of the calendar, urgent to-do, and project lists, you should schedule a specific time each week to conduct a comprehensive review of your workflow. During this time, you can reflect on all of the items demanding your attention and refresh your mind about what is going on in each area of your life. During the weekly review, take the time to update your lists by checking off completed items and moving any information that is no longer an action item to its appropriate place in a project information, reference, or archive file.

The Weekly Workflow Assessment is critical to the success of the entire model. Without regular reviews, you may overlook vital information recorded in the system or lack clarity about your current projects and responsibilities. This is a time to make sure your system is working for you and make any necessary modifications to the model. Without a weekly (and daily) assessment, your information capture system will surely fail.

Make Reviews a Priority

Some of the busiest professionals I know have a hard time with the Assess phase because they're so busy reacting to the incoming stimuli and completing the tasks on their calendars and urgent to-do lists that they never take the time to review and assess the big picture.

You may be able to get by with skipping the occasional daily assessment, but letting more than one day pass without conducting a review is a recipe for trouble. Under no circumstances should the weekly assessment ever be skipped—to do so is to risk destroying the entire Workflow Management Model.

For some leaders, scheduling a time for quiet review is impossible when their teams are present. If simply identifying a time for Assessment is a daunting challenge, you may have to make some concessions in your routine. The Assessment phase is too important to skip, skimp on, or overlook. Try scheduling an hour on Friday evening after your team has gone home for the weekend, or create a place at home where you can conduct your Assessment on Saturday or Sunday morning. No matter how urgent the to-dos and appointments on your lists and calendar may seem to be, your Assess time should be of the highest priority. Without this essential time, the entire Workflow Management Model will eventually cease to work for you.

Act

The final phase of the Workflow Management Model is to begin performing the tasks and activities you've outlined for yourself. The Act phase is the stage you will operate in most frequently—whenever you are not Assembling, Analyzing, Arranging, or Assessing, you will be Acting. The other phases of the Workflow Model exist to facilitate your ability to act and get things done.

Because your brain can trust that it won't forget a critical to-do captured in the Workflow Management Model, your mind should be free of distractions and running at its highest level of productivity.

In Chapter 13, "Project Team Organization," we will return to the Act phase of the Workflow Management Model and explore further how you can tackle the action items on your list with absolute focus and optimum efficiency.

The Workflow Management Model in Action

I once advised the staff of a popular state senator about how to handle the senator's constant barrage of information and better manage his busy calendar. The senator's office received a constant influx of input: e-mails to multiple inboxes; regular postal mail; constant phone calls from constituents who required the senator's assistance in navigating state agencies; invitations to countless ceremonies, events, fundraisers, and parties; and frequent messages from the senator himself. Because the senator prided himself on providing exceptional service and assistance to his colleagues and constituents, each piece of input or correspondence required direct response or action from the staff in order to be resolved.

Simply put, the senator's staff was drowning in all of the communications and information coming at them. With so many different channels for information to come in, simply keeping up with all the input occupied a great deal of time. Responding to or resolving all of the comments, questions, and concerns was another story.

Compounding the problems was the fact that the office was staffed by three part-time interns who were overseen by one full-time office manager. The senator had a habit of handing off instructions to one worker, then asking another employee about the follow-up. When the second employee had no idea what the senator was referring to, he would become frustrated by what he perceived to be a lack of communication among his staff. Constituents would call and speak with one employee about a specific problem, then call back to follow up, only to speak with another worker who knew nothing about the case. Liaisons at state agencies who worked with the senator's staff to resolve constituent issues and complaints would become infuriated by duplicate requests for information from different staffers responding to the same issue.

Information was falling through the cracks left and right, resulting in missed events, angry constituents, and a stressful work environment. The senator's office experienced high turnover as a result of the chaos, and employees burned out after only a few months on the job.

The lack of organization also cost the senator's office opportunities for outreach and growth. Rather than working proactively to plan for events and create strategies to generate positive press, the employees worked frantically around the clock reacting to all of the incoming stimuli. There was little time to plan for any new projects as the staff struggled to even keep up with communications.

Finally, every piece of correspondence sent from the office had to be approved by the senator himself. If a letter or press release was drafted by one employee and the senator did not have time to approve it before the end of the day, the documents would not be sent until the next time the employee was in the office. Other times, the senator would forget to return an item that was sent for his approval, and because there were no follow-up reminders, the item would never be sent at all. In short, many opportunities to make a good impression on constituents and send timely press releases and letters were missed due to lack of organization and communication.

Despite the organizational challenges working against the office, they did have some advantages. Much of the work the staff completed had been done many times before. For example, one of the staff's major responsibilities was to write letters to constituents—congratulations letters, letters of support and recommendation, thank you notes, and follow-ups to requests for information about particular programs or pieces of legislation. Many of these letters were written along the same vein, and could easily have been reused for future use, but the organizational system was so lacking that a new letter was drafted for each occasion instead. Simply put, the chaos in the office's filing system prevented anyone from being able to find what they needed to do their jobs right.

After several days of observing the issues this small office was facing, I introduced the Workflow Management Model to the staff.

Assemble Phase

One of the biggest problems I had observed in the office was the amount of unfinished business occupying each of his staffers' minds. Each employee was on the receiving end of a tremendous amount of input, much of which fell into the unfinished business category. For example, the senator would frequently hand his staffers clips from magazine or newspaper articles he had read and say, "Do a letter about this," or "See what kind of legislation we could develop," without offering further clarification. The staffer would then carry around the newspaper clip in his or her briefcase and worry about how to handle the information, even though he or she had no idea what the desired outcome was or necessary action steps to take.

To offset the tidal wave of information that was constantly sweeping the staff off its feet, I began by introducing a series of collection containers for the staff to use during the Assemble phase of the Workflow Management Model. Rather than having each staffer maintain his or her own list of input they'd received during the day, the office developed a system of collection containers that everyone could access and share. Each staffer maintained a list of the messages he or she received during the day, then e-mailed them to a central inbox in categories of like information. For example, the staffer might e-mail three messages under the heading "Meeting Requests," one message under the heading "Constituent Issues to Resolve," and another five messages titled "Messages for the Senator." All physical copies of information, such as the article clippings that the senator handed to whoever was in his path, went into a central "in basket" on the office manager's desk for processing. The staffer who received the document would attach a sticky note to the document with the date, his or her initials, the source of the document, and any instructions that had accompanied it.

This collection container system enabled the office manager, who was the only person in the office every day of the week, to be aware of all of the information the part-time employees received throughout the course of their shifts. The staffers still addressed any constituent issues or concerns that they could handle in five minutes or less immediately, but they sent all other information and input to the collection basket for processing.

Analyze Phase

Once the office staff completely understood the importance of the Assembly phase, I introduced the next step: Analysis. Whenever his daily schedule permitted it, the office manager was instructed to conduct a complete analysis of the information in the collection containers. Using the Easy Item Analysis system, the office manager would first:

- *Define the item.* Based on the notes from the staffer who took the message, the office manager would determine the importance of the information in order to categorize it in Step 2. Many items in the inbox were predefined thanks to the staffers' new habit of grouping messages in like categories.
- *Determine the action, if any.* If the item did not require a specific action but contained important information, the office manager immediately filed it in a reference file, a future action file, or the appropriate project information file. Each of these files was kept next to his desk for easy sorting. Anything that did not require immediate action was immediately thrown away or recycled. Items that did require a specific action were placed in a separate file for sorting in Step 3.
- *Define the action.* If an item in Step 2 required action, the office manager would now determine how long the action would take to complete. If the task would take five minutes or less, he would complete the task immediately. If the task would take longer than five minutes, he would delegate the item to another staffer to be completed immediately, schedule the task for a future date and time, or make a note of the required activity on the appropriate project list.

Arrange Phase

Upon completion of the collection container Analysis, the office manager learned to Arrange the information and organize a system for ensuring every item he processed was completed. After sorting documents in the Arrange phase, the office manager would have seven piles or files:

- Activities or tasks to be added to an urgent to-do list
- Activities or tasks to be recorded on a project list
- Activities assigned to someone else to be recorded on a pending item list for follow-up
- Items to be scheduled on the calendar for a specific date and time
- Information for a future action file
- Information to be recorded in a project information file
- Information to store in a reference file
- Trash

When the office manager delegated a task to someone else by adding it to one of their project lists or urgent to-do lists, he would record the item on his pending item list for follow-up. This enabled him to keep track of the assignments he was handing out to other staffers and ensure that the task was completed, despite the other employees' fluctuating schedules. Anything that a staffer submitted for the senator's approval was also recorded on a pending item list so that it could be corrected and distributed once the senator reviewed the information. If the senator requested a press release on a new bill, for example, the office manager would assign the press release to an intern and record the assignment on his pending item list. Once the press release was complete, the office manager would then send the release to the senator for approval, again making a notation on the pending item list. When the senator returned the press release with revisions, the office manager would have an intern complete the changes and distribute the release. At that point, the press release could be deleted from all of the office manager's to-do and pending item lists.

The office manager also learned to make the most of the senator's calendar. When the office manager created a new appointment or event to be added to the senator's calendar, he would immediately add all of the critical details about the event to the calendar entry. Contact names and phone numbers for interviews, details about events, and other critical information were added directly into the calendar so that the senator and the office manager had a clearer picture of what needed to happen at each event or appointment.

I then taught the office manager to sort information in the future action file by group. For instance, conferences that the senator was unable to attend but interested in for future years went under the heading "Conferences,

Workshops, and Meetings" in order of date. The senator's ideas for future legislation were recorded under that title, as was information about potential press events. Project information files were stored in a central location next to the project lists so that any employee who needed to review the file could instantly appraise where the project currently stood. Finally, the office manager maintained a detailed reference file of any information that could be used again or referred to in the future. Rather than constantly rewriting the same letters week after week, staffers began filing letters the senator had approved that they were likely going to have to write again. Then, when an occasion that required a letter arose, any employee could turn to the reference file to find a template for the new letter.

Assess Phase

Once everything had been Assembled, Analyzed, and Arranged, the office manager would stop to Assess the current workflow and review each staffer's project and to-do lists. At the beginning of each workday, I taught the office manager to conduct a daily Assessment of the previous day's Analysis.

The office manager would begin each morning by reviewing his calendar, his urgent to-do list, and the to-do lists of each staffer working that day. Then he would turn to the pending item list, and follow up on each task on the list to make certain that it either was still in progress or had been completed. If an item on the pending item list was complete, he would then move it off of that list and record any follow-up actions on the appropriate to-do list or project list, or schedule the follow-up task on the calendar. After ensuring that everything was on track for the day, he would quickly scan the project lists to make sure no tasks were being overlooked.

In addition to the daily assessments, I encouraged the entire staff to gather for a Weekly Workflow Assessment to monitor their progress. During this time, the employees would review the pending item list, calendar, to-do lists, and project lists together and discuss their progress on each item. Then they would turn their assessment to the project information, future action, and reference files to make sure that none of the information in these files had moved from "waiting" to demanding action. Finally, the staff would work together to plan future events, press opportunities, and constituent outreach efforts. Because they were confident that all of the "reactive" input was being properly processed and completed, they were able to become proactive and begin coordinating bigger, better projects.

Act Phase

While the office manager was moving through the Assemble, Analysis, Arrange, and Assess phases of the Workflow Management Model, the other employees were Acting on the tasks they had been assigned. Because a reliable system existed for collecting all of the information they encountered throughout the day, staffers were able to focus on the task at hand and tackle the activities they were assigned more productively.

The introduction of the Workflow Management Model transformed this once-chaotic congressional office into a well-oiled machine. Rather than constantly reacting to stimuli, employees were able to become proactive and develop detailed plans for new projects. Information no longer slipped through the cracks, and every staffer understood all of the issues the office was handling—or, at the very least, knew exactly who was handling the matter so they could refer the senator or callers to the appropriate person for information.

How could your organization or team benefit from the Workflow Management Model?

Final Thoughts

The Workflow Management Model ultimately enables you to focus 100 percent of your attention on the most critical task you must complete at any given moment, while also ensuring that activities and information of a lesser priority are not overlooked and lost. The stages of the Workflow Management Model—Assemble, Analyze, Arrange, Assess, and Act—prepare the playing field for the day. In the next chapter, you'll learn how get your "players" ready for the game as we shift our attention to Project Team Organization.

For a diagram and other useful materials on the Workflow Management Model, please see our companion website at www.ManagementToLeadership.com.

Chapter 12 Review

Project leaders must be able to "get in the zone" to work at the highest levels of productivity and concentration. Getting in the zone requires that the project leader have a system for capturing unfinished business and other information and input encountered throughout the day so that he or she can focus on the task at hand with maximum concentration. The Workflow Management Model is the project leader's answer to this challenge.

There are five phases of the Workflow Management Model:

- Assemble
- Analyze
- Arrange

- Assess
- Act

Goals of the Assemble Phase

In the Assemble phase, project leaders learn to collect the input into collection containers for future processing. Anything in need of attention or action should be physically stored in a collection container (or represented in the container in some way).

- Collect every piece of information and input in need of attention in a safe and reliable collection container.
- Store these items until the Analyze phase, when they will be processed.
- Use as few collection containers as possible.
- Keep the collection containers accessible at all times.
- Clear out the collection containers on a regular basis.

Goals of the Analyze Phase

In the Analyze stage, project leaders process the input gathered in collection containers. During this phase, project leaders follow a four-step process to determine the next step for each item. If no action exists for an item, it is either thrown away or stored in a future action, reference, or project information file.

- Nothing goes back into the collection container.
- Any tasks that take less than five minutes are completed.
- Nonurgent tasks are delegated to a team member or scheduled for completion at a future date.
- Everything in need of action is prepared for processing in the Arrange phase.

Goals of the Arrange Phase

In the Arrange phase of the Workflow Management Model, project leaders sort each of the input items into one of eight categories: the trash, future action files, project information files, reference files, urgent to-do lists, project lists, pending item lists, or a calendar. Each item is recorded in the appropriate place for future assessment and activity.

- Every item is scheduled, recorded, or delegated to ensure that it is completed.

Goals of the Assess Phase

In the Assess stage, project leaders review the processes and items that they've planned to get a firm grasp on what needs to be done. By the time a project leader arrives at this phase, he or she should have an up-to-date to-do list, calendar, and project lists to review.

- Conduct daily assessments of the to-do list and calendar.
- Conduct weekly workflow assessments to review the entire model, including information stored in pending item lists, project information files, and future action files.

Goals of the Act Phase

In the Act phase, processing and planning are finished. At this point, project leaders begin performing the tasks and activities they've outlined to complete—this is the actual "work" phase of the Workflow Management Model.

- If the Workflow Management Model has been successfully implemented, project leaders should be able to work on the tasks they've outlined with optimal concentration and efficiency. In the Act phase, project leaders should be working at a maximum level of productivity.

13

Project Team Organization

Your team's performance will ultimately determine your success as a project leader. No matter how efficient you are at getting things done, you will be unable to complete everything in need of attention without the assistance of your team. The Workflow Management Model requires that you delegate tasks to your team, but the organizational system will not help you if the things you assign to others are done wrong or not completed at all.

You must be organized and efficient in order to lead your team; however, your team's organizational skills are of equal (perhaps even greater) importance. You must be able to rely on members of your team to complete tasks you delegate to them promptly and effectively. Team members must be cooperating and coordinating on specific tasks and throughout the entire project, while exemplifying the team's mission, upholding the team's values, and working toward a successful outcome for the project as a whole.

Achieving Organizational Buy-In

When an already high-performing team embraces the Workflow Management Model of getting things done, the results are astounding. People are amazed at the change in a team's efficiency when individual team members are able to "get in the zone" and focus on each activity without distraction.

It's easy to see the differences between teams that have completely embraced organizational systems and teams that have not. Organizations with complete buy-in for some form of workflow management model are more focused, more in synch, and more productive. When the team truly believes that all of the information, input, and correspondence are being captured and nothing is being overlooked, they will achieve higher levels of productivity in taking on specific tasks, which enables them to achieve a successful outcome on specific projects faster and with fewer headaches.

The Workflow Management Model should be part of every team member's daily routine; however, some variation in the specific implementation is to be expected. You cannot dictate the way someone manages his or her personal space, but you can make your team accountable for whether or not things are getting done as they should. Every member of your team should be following the Assemble, Analyze, Arrange, Assess, and Act process of filtering input and information. Beyond that requirement, the way team members manage their to-do lists and personal calendars should be left up to each individual's discretion, so long as their lack of organization isn't affecting other team members' performance.

Handling Gaps in the Workflow Model

The project leader's role in the Workflow Management Model is to set an example for his or her team and ensure that there are no gaps in the system. For example, if one person on your team is not regularly Analyzing the items in his collection containers, he may be preventing other team members from completing key tasks on their agendas and holding up the rest of the team's progress.

The problem becomes exacerbated when important players on the team fail to respond quickly to information in their collection systems. Imagine if a senior member of the team all but ignores messages left on her voicemail by failing to record them in a collection container for Analysis. Any critical messages left by another team member regarding an item that he is working on will be left unprocessed, which will then prevent that team member from moving forward on his next task.

Identifying and Solving Issues before the System Fails

Spotting and addressing problems with any team member's ability to follow through on the Workflow Management Model is critical to keeping the system afloat. If one person on the team is not responding appropriately to the items placed in his or her collection containers, this can cause other team members to lose faith in the system and abandon their organizational system as well.

Fortunately, these problems should not continue for long. To ensure that the system continues to run smoothly:

- Persuade team members to maintain and update their pending item list to reflect anything they've delegated or shared with another team member, and then periodically follow up with the other party. If team members are maintaining and frequently updating their pending item lists, they should notice any gaps in the system quickly.
- Encourage team members to share any concerns they may have with another team member's responsiveness with you directly. As the

project leader, you can then address the guilty party directly and help them get back on track.

As I've said before, the best project leaders delegate results—not processes. Hold your team accountable for getting things done, but trust them to manage the minutiae appropriately.

The Project Team Decision-Making Process

The project leader is charged with helping team members to understand how they will collaborate on each task within the bigger project to get things done and achieve the desired results. This requires the team to make decisions, on both the individual and group levels, on where to spend their time and how to best move a project forward.

In the course of researching and writing this book, I realized that there are many decision-making processes for teams, none of which are actually project management driven. The following Project Team Decision-Making Process is a good example of how teams can make decisions that move projects forward in a structured format. Teams need to be able to evaluate what needs to be done, what tools are available, and how the task or project fits within the bigger picture in order to move forward quickly and efficiently.

Defining a specific process by which project teams make decisions will allow them to evaluate the ideas and select the best path forward without immediately jumping to "Let's get it done." The Workflow Management Model should provide your teams with a clear picture of everything that requires attention or action—the Project Team Decision-Making Process will help them determine what to work on when and shape the direction of projects as a whole.

The phases of the Project Team Decision-Making Process are as follows:

- Understand
- Brainstorm
- Investigate
- Plan
- Execute
- Stop and renew
- Resume
- Review

Let's explore each stage of the process in greater detail.

Understand

To maintain balance within the team triangle, you must show the team why they are completing certain tasks so that they can be sure the outcome of the task fulfills the intent of doing the activity in the first place. The Understand phase is designed to give the entire team the ability to comprehend what things need to be done and the criteria by which to judge when the task is complete. In the first phase of the process, two things must happen:

- The project team must understand and clarify what needs to be done. What is the team doing, and why are they doing it?
- The success criteria for the project must be defined. What is the desired outcome, and how will the team know when they've achieved it?

At the conclusion of this phase, teams should know why they're working on the project and what a successful outcome will look like.

The First Question You Should Always Ask

> The person who knows "how" will always have a job.
> The person who knows "why" will always be his boss.
>
> **—Diane Ravitch**

One of the most important questions you can ask before embarking on a new task, project, or activity is "Why?" Despite childhood warnings against questioning the purpose of a given command ("I'm your mother, I don't have to tell you why"), considering the purpose of any task is critical. After all, there's no point in completing a task if the end result doesn't fulfill the original intent.

For example, imagine you ask a colleague (we'll call him John) to purchase a camera for the team to use at an upcoming company retreat. You plan to use the camera to take videos to upload on your company's website when you return, but you don't tell John why you want the camera. You just ask him to go pick up a camera and bring it with him to the meeting. John knows that you normally aim to purchase the least expensive supplies possible, so he picks up a pair of disposable cameras and brings them to the retreat. The disposable camera has no video-recording capabilities, and thus is of no use to you in terms of your original intention. John did exactly what you asked, in accordance with your normal expectations, yet the outcome is incorrect and John's efforts were pointless. If John had only known why you wanted the camera, you would have received a tool that fulfilled your needs.

Considering the "why" of any activity is beneficial to teams for a number of reasons:

- *It provides criteria for evaluating options.* If one option will not meet the intention, it can be eliminated as a choice. For example, had John known why you needed the camera, a disposable camera wouldn't have even been a consideration. You need to know what you are trying to achieve in order to make the best decisions possible. If, for instance, you are unsure how much to spend on marketing a new position, consider the value of that position. Obviously, the role of CEO requires you to attract a higher caliber of talent than a maintenance staff member. Considering why you are doing a task in the first place helps define the best options for getting the job done.

- *It helps develop focus.* When everyone understands why they are doing something, confusion is eliminated. Knowing why something must be done helps people clarify what they are trying to accomplish and eliminates the need to spend time considering strategies or options that will not achieve the goal.

- *It builds motivation.* People like to know that they are doing something for a reason. If they perceive a task as meaningless, unimportant busy work, they won't be inspired to get the job done because they don't believe the task is significant. On the other hand, understanding how an activity will move the project toward completion is an incentive for getting the task done.

- *It defines success criteria.* People prefer a triumph over a failure. When the team knows why they are doing something, they are better able to evaluate which strategies will best fulfill that purpose and which will not. If John knew why he was getting the camera, he would have known which type of camera would be considered a successful purchase.

Focusing on the Desired Final Outcome

Keeping the ideal conclusion of a project forefront is essential to being a smart and efficient team. Asking "why" a task, project, or activity must be done is the first step in focusing on the desired outcome, but you must describe the desired outcome beyond answering "Why?" It's difficult to choose the right activities to perform or the right places to spend your time or money if you are unsure where the project is ultimately supposed to be heading.

In the Understand phase, your team must understand what they are working toward so that they can best allocate the team's time and resources. What does a successful outcome look like? How much does it cost? How quickly is the end result reached? Who is responsible for designating the task or project as complete? Each of these questions must be understood if the team wants to work effectively and efficiently toward the finish line.

Every activity or task the team undertakes should be evaluated from the perspective of outcome: does this activity help move us closer to the desired result? If the answer to that question is no, the activity or task could very well be scrapped altogether. If the answer is yes, the team must determine why the task moves them closer to the goal and identify the outcome that will fulfill that purpose.

Once your team understands why they are doing a task and what the desired outcome looks like, they can move on to the Brainstorming phase of the Project Team Decision-Making Model to begin generating strategies and ideas for executing their plans.

Brainstorm

When your team understands the purpose of a project or activity and has defined the desired outcome, they can begin considering the many options for getting it done. This step can be skipped in some decision-making sessions since brainstorming may not be necessary for every task you consider. For example, there's no need to brainstorm about how John will go purchase a camera for the retreat—that is an obvious task with a clear strategy for completion. The decision-making process at the onset of a new project, on the other hand, could certainly benefit from a good brainstorming session. Just because the first idea a team has is a good one doesn't mean it's the best idea possible. Brainstorming is a great way to uncover new methods of getting things done, especially when the entire team is clear on the purpose and desired outcome of the task or project.

While brainstorming should not be a formal or complex process, having some ground rules in place can help breed better ideas. The following brainstorming "rules" will help your team achieve more productive brainstorming sessions that generate better ideas and outcomes.

Brainstorming Rules

1. Assign one team member to write down ideas for the group.
2. The rule list and brainstorming list must be displayed in clear view where everyone can see.
3. The person recording the ideas must write down all of the suggestions, even if an idea is considered outrageous or ridiculous.
4. Invite at least two people who are not members of the team to participate in the brainstorming session to get a fresh perspective.
5. Encourage participants to come up with as many ideas as they can during the time dedicated to brainstorming. The more ideas generated during the session, the better.

6. No one is allowed to elaborate on his or her idea during the brainstorming session, except to elaborate when an idea is not immediately clear.

7. No one is allowed to judge or ridicule another person's idea—leave your opinions at the door!

8. Everyone is considered equal during the brainstorming session—the project leader has no more seniority than any other team member.

9. Encourage team members to "snowball" on other ideas thrown out during the brainstorming session.

10. The brainstorming session must continue until the allotted time has expired—you never know when genius might strike!

Brainstorming in this way is a powerful tool for stimulating your team's innovation and creativity. Because the notion of brainstorming rules is a new concept to many, let's discuss these guidelines in greater detail.

Rule 1: Assign One Team Member to Write Down Ideas for the Group

In many cases, the ideas generated during the session will not be the solution, but rather a starting point toward a solution. During every brainstorming session, one team member should be the designated recorder. That way the team has a written record of the ideas to discuss when the actual "storming" of ideas is complete.

Rule 2: The Rule List and Brainstorming List Must Be Displayed in Clear View Where Everyone Can See

Having the list clearly displayed where all team members can see it will help stimulate more ideas and ensure that the recorder is actually recording every idea. Posting the rules in clear view ensures that they are followed.

Rule 3: The Person Recording the Ideas Must Write Down All of the Suggestions, Even If an Idea Is Considered Outrageous or Ridiculous

The person recording the team's idea is NOT permitted to censor the team's creativity by omitting suggestions that he or she views as absurd or unproductive. Displaying the brainstorming list in clear view ensures that this rule is followed.

Rule 4: Invite at Least Two People Who Are Not Members of the Team to Participate in the Brainstorming Session to Get a Fresh Perspective

While this idea might seem foreign to some, inviting two non–team members is a great way to infuse fresh ideas into the brainstorming session. Preferably, these two guests should be from a completely nonrelated area of the organization or from an entirely different company.

*Rule 5: Encourage Participants to Come Up with as Many Ideas
as They Can during Time Dedicated to Brainstorming*

The more ideas generated during the session, the better. Do not stop coming up with new ideas before the allotted time for the session has expired. You never know when the best idea of all might pop into someone's mind.

*Rule 6: No One Is Allowed to Elaborate on His or Her Idea during the Brainstorming
Session, Except to Elaborate When an Idea Is Not Immediately Clear*

The brainstorming session should be a rapid-fire flow of ideas. Do not allow this creative flow to be stifled by permitting team members to go on and on about a specific idea. It's OK to offer a sentence or two for clarification, but there should be no discussion about a specific idea until the conclusion of the brainstorming session.

*Rule 7: No One Is Allowed to Judge or Ridicule Another Person's Idea—
Leave Your Opinions at the Door!*

The rule prohibiting discussion of ideas should prevent on-the-spot judgment; this is an important point. Team members should feel encouraged to express their creativity and throw out crazy ideas during a brainstorming session. At the first hint of judgment or ridicule, the group's willingness to throw out new ideas will be hindered. Do not permit judgment to stifle the team's brainstorming session.

*Rule 8: Everyone Is Considered Equal during the Brainstorming Session—
the Project Leader Has No More Seniority Than Any Other Team Member*

This can be a challenging rule for some project leaders, but during the brainstorming session, all team members are equal. The team leader's or boss's ideas get no more emphasis than another member of the team's ideas.

*Rule 9: Encourage Team Members to "Snowball" on Other Ideas Thrown Out
during the Brainstorming Session*

Sometimes, the best ideas are triggered by another suggestion thrown out by the team. This kind of snowballing should be encouraged, as many of the best ideas will be inspired by someone else's thoughts. Having the brainstorming list posted in clear view will help encourage this kind of collaboration.

*Rule 10: The Brainstorming Session Must Continue until the Allotted Time
Has Expired—You Never Know When Genius Might Strike!*

No matter how great the ideas on the board may already be, ending the brainstorming session early is not permitted! Encourage your team to continue throwing out ideas until the session time runs out—the best idea might spring into someone's mind in the last seconds of the session!

Brainstorming hint: This whole process should be done very quickly. It is not designed to be a slow and laborious process. It should be very quick action that keeps people focused and helps them generate ideas without thinking about what other people would say.

Investigate

Once you understand what you need to do, you must evaluate what you have available to do it. During this phase, the team must look at not only the resources that it has, but also the resources it will need to have. During the Investigate portion of the process, the team may realize it needs more people, such as subject matter experts, or more resources, such as funding.

In the Investigate stage, teams should consider the following:

- Where does the project stand right now?
- What do we need more of?
- Is obtaining these additional resources a possibility?
- If not, how can we make do without the extra tools?

Using Available Tools and Resources

If your team adheres to the Workflow Management Model described in the previous chapter, they should have two very important tools available to aid in the Investigate phase: project information files and reference files.

Project Information Files

Your team's project information files are an excellent resource during the Investigate phase of the Project Team Decision-Making Process. The project information files should contain any research, data, details, plans, notes, documents, or other information pertaining to any upcoming or ongoing projects. These files should contain everything that the team needs to know to execute a specific project, and must be kept up to date with the latest information or developments regarding a specific project. The project information file for a specific project should reflect the most up-to-date and comprehensive picture of where a project stands at any given moment.

If a project information file has not yet been developed for a new project, the results of the Investigate phase of the process should be used as the foundation for this file. Any information gathered during the decision-making process should be included, as well as the results of the team's decision regarding the project.

Reference Files

In addition to the project information file, your team should have access to general information files where ideas, information, or other data that do not directly correlate to existing projects or plans are recorded. The reference files should be broken down by category so that information is readily accessible when your team needs it. Any information that your team uncovers during the Decision-Making Process that might be useful for another project should be recorded in the reference files for future use.

If your team takes the time to develop the following resource, it will be a valuable tool in the Project Team Decision-Making Process.

Past Project Histories

One of the most powerful resources your team can use during the Investigate phase is your past project history information. I encourage teams to keep not only records of how they plan to execute any project, but also a final report on how the project was actually implemented. At the conclusion of any large project, the team leader (or a designated team member) should record information about the actual execution of the project to determine three main things: what went right, what went wrong, and what could be done better next time. When drafting the record, the person should consider the following:

- Where did the team succeed?
- Where did the team make mistakes?
- What were the risks going into the project? Were those risks realized?
- Were there any unexpected roadblocks?
- Were there unexpected victories or processes that were easier than expected?

Having a record of how a project was actually executed is a beneficial tool when making decisions about future projects. A written record ensures that the team will not make the same mistakes twice, and past precedent may offer insight on the best way to proceed in an uncertain situation.

Before moving on to the next step of the Decision-Making Model, and actually Planning how they will execute a task or activity, team members should understand what existing resources and technology they can leverage to make their job easier.

Plan

> In preparing for battle I have always found that plans are useless, but planning is indispensable.
>
> **—President Dwight D. Eisenhower**

All great project managers must spend time in the planning phase. It is vital you spend time reviewing and integrating what you've learned in the previous two phases to ensure that you have a plan that will meet not only your success criteria, but also those of your customers.

Before moving to the Execute phase, everyone on the team should understand what the "action steps" of the project will be, and which tasks they are accountable for completing (and in what order).

Mission Critical: Define the Next Action

Without a doubt, the most important conclusion that any planning sessions should come to is the agreed-upon definition of the group's next action. Making it a priority to define the next action before the planning meeting adjourns will increase the efficiency of your team in ways that you can't imagine. I've sat through a number of meetings where the team talks for forty-five minutes to an hour about huge, vague plans and everything that needs to happen before the project's deadline. I'll hear mention of enormous tasks that could never be tackled straight on. Rather than breaking these activities into smaller action steps and delineating them to individual team members, everyone treats the giant task with the sense that it will just get done because everyone is "working together." After the team brings up all these vague to-dos, the meeting is abruptly adjourned and everyone rushes back to their offices, without knowing what they should really be working on next or who is responsible for accomplishing what activity.

Asking the question "What do we need to do to move forward?" at the conclusion of every planning session (or, better yet, every conversation!) improves performance because this question shifts the way your brain frames the meeting. Asking this question stimulates the final round of questions necessary to make real decisions about how to keep the project moving. Just as asking "why" you are doing something is critical to ensuring that the end result fulfills the purpose, defining "what" makes certain that the project moves forward and every team member understands the next actions.

Defining specifically what the team will do next solves several common problems that groups face at the end of meetings. Deciding upon a next action does the following:

- Establishes a natural conclusion
- Breeds responsibility
- Encourages action
- Empowers the team

In Conclusion …

Defining "what happens next" steers the discussion toward a meeting summary. After the team has hashed out what it has learned and discussed the various options for proceeding, this question brings the group to a decision point. The group has to ask itself, "OK, based on what we just learned, which path are we taking, and, more importantly, exactly what is each of us doing to move this forward?"

Too many meetings adjourn in such a manner that the participants have no idea what was actually decided. This often breeds the belief that meetings are pointless or a waste of time. In other cases, the general plan is agreed upon, but the specifics of how to move forward are left up in the air. This creates unfinished business to occupy everyone's minds, whereas specific action assignments would have created to-dos that move the group closer to the goal.

Realizing Responsibility

Another benefit of defining the next action is the clear delineation of responsibility amongst team members. When the question "What is the next action step?" is asked, the team needs to be prepared to not only define that action, but also assign a person who will be accountable for completing the task. Next to the specific action to be performed, the meeting note taker should record the name of the person responsible for completing the task. This assignment should then be recorded on a pending item list for follow-up in the future.

Encouraging Action

When the group has arrived at this point in the decision-making process, they have already evaluated their options and examined their available resources. By defining the next action, you allow the team to go ahead and act.

Encouraging action is especially crucial to projects with distant deadlines. Sometimes, projects that are too far off in the future are hard to keep moving. Yet just because something is not due immediately doesn't excuse a complete lack of progress. By continuously defining a next action for long-term projects, you'll keep these projects moving toward completion.

Empowering the Team

The final benefit of deciding the next action is the sense of empowerment that it grants the team. When people know what needs to be done next and understand how completing that task fits into the larger goal, they gain control of the situation. Too many times, employees are unable to move forward on projects on their own and are reliant on someone else continually coaching them and outlining the next step. When you begin forcing people to define the next action, they will gain the sense of control and accomplishment that comes with getting ahead of the game.

Execute

Step 4 of the decision-making process is to actually make the work happen. Most people start the project at this phase, and consider the previous three steps when problems arise. By the time you reach the execution phase, every member of your team should know exactly what they are accountable for doing, why they are doing it, what the expectations are, and what a successful outcome will look like.

Stop and Renew

This step is a bit unusual. As you move through the execution process, it is necessary to pause or stop at certain parts to reflect on your progress. As you're moving through the project, stop and evaluate what you are doing, determine what you are doing well, and identify places for improvement. It's essential to reflect and ask yourself several important questions:

- What should we do more of? (What are we doing now that works?)
- What should we do less of? (What actions or processes are stifling productivity?)
- What should we start doing? (Is there something we could be doing now to make things move faster?)
- What should we stop doing? (Is there an unnecessary or counterproductive activity that we could drop altogether?)

The Stop and Renew phase offers a great opportunity for the project team to gather and reflect upon the progress that has been made and prepare for the work that remains to be done. Asking these questions will not only improve your team's performance on the project that they are currently working on, but also offer valuable insight into how to approach future projects. Figure 13.1 offers examples of what types of suggestions your team might use during the Stop and Renew phase of the decision-making process.

Once you've identified what you should start (and stop) doing and examined your progress, continue executing your process, implementing the necessary changes.

Resume

The brief assessment in the previous step should serve as a good renewal for the team, and will help make your future progress run smoother. It doesn't

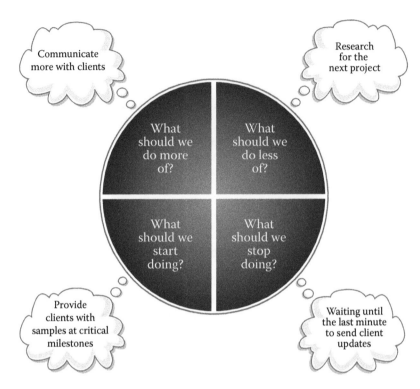

FIGURE 13.1
The question circle.

hurt to periodically stop, review, and renew again to make sure you're continuing in the right direction. As the project leader, you must be actively involved with the team throughout the execution of the project, working with the team and leading from behind.

Review

When you have completed the project, it is essential to go back and look at what you've learned throughout the process. This will help you provide your team with good feedback, which will in turn force them to consider how they would do things differently and improve the way they work together in the future.

The review process should include recording details about the project's execution, like the project history described in the Investigate phase of the plan. Be sure to record what went right, what went wrong, and what went as expected. Write down details about anything that presented an unforeseen obstacle or any challenges that your team had not considered in the decision-making process. Having records to look back on will improve the efficiency of your decision-making process in the future and help your team grow.

Conclusion: The Questions You Ask Are Key

Increasing the efficiency of your group doesn't require a tremendous transformation. You can dramatically increase your team's efficiency by implementing simple, logical methods and processes that will streamline the efforts of the group.

For example, asking what needs to happen next and understanding why your team is completing a specific task are easy ways to clarify the objectives and achieve the best results. If you don't take the time to ask, "What needs to happen next?" or "Why are we hiring an account executive?" or "What is the purpose of starting this program?" how can you make the best decisions? You must ask the right questions to ensure that your team's actions actually move the group toward its goals.

Chapter 13 Review

In order for the Workflow Management Model to succeed, every member of a project team must buy in to the system. Organizations that wholly embrace the model are more focused and more in synch, and achieve higher levels of productivity. When teams trust the system and know that all of the tasks that need to be tackled are being captured, they can focus on the work in front of them and get more done faster.

Every member of the project team must embrace the Workflow Management Model, but variation in implementation is acceptable. So long as every member of the team is able to capture, process, and filter the critical information they encounter in a way that ensures every item is completed, minor variances in implementation are perfectly fine. Remember: team leaders delegate results, not processes.

The Project Team Decision-Making Process

Teams absolutely need a decision-making process that enables them to evaluate what needs to be done, what tools are available, and how the task or project fits within the bigger picture. A universal decision-making process will help the team make better decisions and enable them to get things done faster.

The phases of the Project Team Decision-Making Process are listed next.

Understand

In this phase, project teams must understand why they are completing the task so that the outcome fulfills the purpose of the task. At the conclusion of this phase, the team should be on the same page about what tasks need to be done and the criteria they must meet to achieve a successful outcome.

Brainstorm

During the Brainstorm phase, teams explore the options for getting things done. Brainstorming may not always be necessary; however, taking the time to explore a variety of ideas is a great way to uncover new ways to get things done and build the team's motivation. To get the best benefits from each brainstorming session, project teams should abide by the brainstorming rules.

Investigate

The Investigate phase of the decision-making process is critical to choosing the best course of action. During this phase, team members should examine the resources they have available, and consider what additional resources they may need. Team members can use the project information files, reference files, and past project histories as valuable tools in this stage.

Plan

The Plan phase is an opportunity for the project team to review what they've learned in the previous phases. At this point, the team should determine whether each possible plan will meet the success criteria for the task and whether the appropriate resources exist. At the conclusion of the Plan phase, every member of the team should understand what the action steps of the project will be and what tasks each team member is accountable for completing.

Execute

When the team reaches the Execute phase, they should be performing the action steps outlined in the Plan phase. At this point in the process, every member of the team should know what they are doing, why they are doing it, what the guidelines for completing the task are, and what a successful outcome will look like.

Stop and Renew

The Stop and Renew phase is the most unique stage of the Project Team Decision-Making Process. This step gives project teams an opportunity to reflect on their progress and ensure that they are headed in the right direction. This is also an opportunity to determine whether the team could be performing at a higher level. Teams should ask themselves four questions:

- What should we do more of?
- What should we do less of?
- What should we start doing?
- What should we stop doing?

Resume

After the team has completed its personal assessment in the previous phase, they should resume the work, adjusting their performance according to the results of the Stop and Renew discussion.

Review

When the team reaches the project finish line, it is critical that the project leader and project team review their performance and determine what they've learned throughout the process. A team historian should record details about the project's execution, including what went right, what went wrong, and what went as expected. Preparing these records of past performance will improve the team's ability to make and execute better decisions in the future.

14

Taking on Tasks

When you break "getting things done" down to its lowest common denominator, the tasks are what propel you toward the finish line. The activities you and your team undertake each day determine your level of success. The tasks you tackle are the sole engine driving your projects forward or running it off course. The decisions you make regarding what you work on and when, and what you choose to delegate or drop altogether, are the most important choices you'll make in "getting things done."

Tasks and the Team Triangle

As you'll recall, the team triangle is composed of three levels: the individual, the team, and the task. Each of these levels works in unison to determine the overall strength of the team. The team, composed of individual members and led by a project manager, must approach each task with the balance of control and autonomy required for effective results. Every team must excel at the task, team, and individual levels to accelerate its performance. If one side falters, the entire team (and perhaps the entire organization) is affected.

Let's recap the criteria for a successful task side of the team triangle.

The What, Why, and How of the Task Is Defined

Before any team member starts a task, he or she should know the "what," the "why," and the "how." In other words:

- What is the task? (What are we doing?)
- What is the expected outcome of the task (Why are we doing it?)
- What are the guidelines for completing the task? (How are we supposed to get it done?)

Team members absolutely must comprehend the expected results from the activity and the guidelines for performance. What are the expectations for quality level? Are there ethical issues to consider? What activities are the highest priority for this project?

Once these initial criteria have been clarified, the team will need to put the task in the greater context of the project:

- What needs to be done now, next, and in the future to complete this task?
- Who is responsible for each step?
- What happens when this task is finished?
- How does the team move onto the next task when this activity is complete?

These guidelines and expectations reflect how the task fits within the bigger project and contributes to the outcome. Team members should understand their specific roles and responsibilities within the given task, including the expectations for quality requirements. Members should also know what resources and technology they have available to leverage the effectiveness of the task from the beginning. All of the groundwork should be laid so that team members are free to move from getting ready to getting things done.

In this chapter, we'll focus on taking on tasks. You'll learn the following:

- How to complete tasks most efficiently
- How to determine what activities to work on, and when
- How to use a calendar to track the tasks and activities you need to complete

Batching for Increased Efficiency

We've already talked a great deal about how to work at the highest levels of productivity. In Chapter 12, you learned how to categorize unfinished business and clear your mind so that you can focus on each task with optimal concentration and focus. However, there are other ways that you can improve your efficiency and learn to complete tasks faster without compromising quality.

The best approach to taking on tasks is a system that I refer to as "batching." The premise of batching is quite simple: in order to achieve the greatest speed and efficiency in your work, you tackle like tasks at the same time.

Batching is effective because it caters to the way our brains operate. The human mind works best when allowed to concentrate on a single type of activity. When we work on a task that is repetitive in nature and does not require the brain to learn anything new or solve a problem, our brain can turn on auto-pilot mode and speed through the activity with ease. However, if we constantly interrupt our brain's activity by switching our focus from one activity to another or challenging the brain to think in new ways, that productivity will be stifled.

By batching tasks in a way that is favorable to your brain, you can get more done faster. You can batch tasks in a number of different ways: you can batch like processes within one task, or you can batch like tasks by scheduling them in the same time block.

Assembly Line Batching

A great example of batching like activities within one task would be an assembly line. If you think back to the Industrial Revolution, the assembly line method of manufacturing introduced by the Ford Motor Company in the early 1900s was considered a major breakthrough that forever changed mass production.

Imagine that your team's assignment is to manufacture 200 dolls. Rather than having one worker assemble one doll at a time and then move on to build the next doll, you could create an assembly line process to take advantage of the human brain's preference for batching. One worker might just fasten each doll's arms and legs to their bodies all day. Then the next worker might paint on the doll's face. The next employee on the line might be responsible for dressing the doll, while the next person puts on the shoes. Each person on the assembly line stays focused on one repetitive activity throughout the day, which enables them to work at the highest level of productivity. Of course, you can apply the assembly line mentality to your work even when there is only one person working.

Here's another example: imagine if you were writing a series of reports for your employer. Rather than sitting down and writing one report at a time, you could form your own assembly line to enable you to work faster. You might start by outlining each report, then conducting research, writing each report, and finally moving on to the editing phase. Rather than constantly switching gears back and forth between outlining, researching, writing, and editing, batching out the tasks in this way would enable you to complete each step faster, ultimately decreasing the time it takes to write all of the reports.

To start batching like tasks for a group of activities, begin by writing out the steps for each activity. Then, group like tasks together. For instance, you might make all of the phone calls for your daily projects at one time, and do the research for a variety of projects all at once in another block of time. Because your tasks and daily activities are outlined on your to-do lists and calendar, you should be able to identify tasks with like activities that can be batched with ease.

Batching by Scheduling

Batching by scheduling builds upon the principles described in assembly line batching. This type of batching mentality should be applied during the development of your daily calendar. You might work for a period of time on one set of tasks that relates to work at the computer, while another block of time is dedicated to phone calls. Your meeting schedule for the day might comprise one batch of activities, while your errand list might form another batch of like tasks.

For example, your daily routine might look something like this:

- 8:00 a.m. to 8:30 a.m..: Review calendar and daily to-do lists
- 8:30 a.m. to 9:30 a.m.: Work on highest-priority to-dos
- 9:30 a.m. to 10:00 a.m.: Sort e-mails
- 10:30 a.m. to 11:00 a.m.: Respond to high-priority e-mails
- 11:00 a.m. to 12:00 p.m.: Make phone calls scheduled for that day
- 12:00 p.m. to 1:00 p.m.: Lunch
- 1:30 p.m. to 2:30 p.m.: Open office hours for team member visits
- 2:30 p.m. to 4:00 p.m.: Work on Category 2 to-dos
- 4:00 p.m. to 4:30 p.m.: Listen to phone messages
- 4:30 p.m. to 5:30 p.m.: Daily processing of collection containers, and calendar and to-do list review

As you can see in the sample calendar above, like activities are blocked at the same time. This person is not switching back and forth between e-mails, errands, and daily appointments, but budgeting an allotment of time for each activity. There is still some flexibility in this schedule, as in the times allotted for taking on high-priority and other daily to-dos, but even open times are batched by activity type.

Batching is essential to performing the most work in the shortest period of time. By playing up your brain's preference for focusing on one task at a time, you can greatly reduce the time it would normally take to finish a task or a set of tasks.

Multitasking Is a Myth

There is a pervasive misconception in our society that multitasking is a productive behavior. The logic goes something like this: by working on more than one task at once, we're getting more things done at the same time. "If I can check my e-mail, catch up on the news, and listen in on this conference call at the same time, I'm getting all three of these things finished in the time it would take to do just one thing."

The reality is that multitasking is a myth. People who multitask are actually less efficient and productive than those who focus on one task or activity at a time. David E. Meyer, director of the Brain, Cognition and Action Laboratory at the University of Michigan, said that working on more than one task at a time is actually detrimental to our ability to get things done—especially in terms of getting things done right. "Multitasking is going to slow you down, increasing the chance of mistakes. Disruptions and interruptions are a bad deal from the standpoint of our ability to process information."

A study out of Vanderbilt University's Human Information Processing Laboratory demonstrates that working on two tasks at once really does slow people down. Study participants were asked to perform basic sound and recognition tasks while the researchers asked questions. When participants were instructed to perform two tasks and then asked a question, their responses were delayed by one to two seconds as opposed to the participants only performing one task at a time.

People aren't really "multitasking" when working on two or more activities at once. Rather than accomplishing multiple things simultaneously, they are simply switching back and forth between two different activities—and they aren't working particularly quickly or efficiently.

A study at Microsoft revealed more surprising news about the effects of multitasking: disruptions cost time over and beyond the productivity lost due to the break in focus and concentration on your work. The research showed that if an employee was interrupted by an e-mail, phone call, or chat message, nearly fifteen minutes passed before the employee resumed the task he or she had been working on. Employees do not "multitask" when they stop to check their e-mail during the day; they actually stop working—and they don't start back up immediately. While fifteen minutes might seem like a negligible amount of time, do the math and calculate how much fifteen minutes cost when considering every e-mail or phone call received by every employee in an organization each day. Now consider the amount of time lost in a week, month, and year. That is an incredible amount of lost time with absolutely no gain!

Multitasking is not only a myth but also a drain on workplace productivity. Project leaders should encourage their teams to focus on one task at a time for optimum efficiency and productivity.

Choosing What to Do on Any Given Day

How you choose to spend your workday will determine whether or not you move closer to your goals and/or the successful completion of a project. Throughout the course of your day, you'll likely be focused on one of three activities:

- Tackling reactive or otherwise unexpected, high-priority activities
- Completing the tasks that you have calendared for a specific time
- Processing your input and scheduling your day through the Work-flow Management Model

While the second and third types of activities are the most essential to moving your team forward on a project, most leaders are bogged down by the reactive and unexpected tasks they encounter throughout the workday.

Reining in Reactive Work

Reactive work has a way of limiting your efficiency in the office. E-mail, phone calls, voice-mail, texts, and human visitors are considered reactive because they require you to react or respond to a stimulus and prevent you from focusing on the proactive tasks and activities you must complete in a given day.

I strongly encourage leaders to schedule specific times of day for these kinds of activities. By reining in reactive work and isolating it to a predefined period of time, you can be more productive during the rest of the day. Calendaring a specific time for responding to stimulus requires less time and mental exertion because you aren't constantly switching gears and moving from one activity to another. As you've already learned, multitasking is a myth. When you attempt to respond to e-mail while working on another activity, you aren't getting both tasks done at the same time; you're simply switching back and forth from one activity to another. Keeping your brain focused on similar activities allows for optimum concentration and activity.

Accept phone calls from clients and colleagues only at specific times of day. There is no reason for you to answer the phone every time it rings; in fact, doing so will prevent you from completing the activity you have calendared for that time. If a call is not scheduled on your calendar, you should allow your voice-mail or assistant to take a message for you. Record a message (or write a script for your assistant) that tells callers that someone will return their call within the next business day and provide a number for them to call in case of an emergency. Then dedicate specific times of the day to reviewing and responding to your messages. For maximum efficiency, request callers to provide the best time for them to receive a follow-up phone call. This will help eliminate back-and-forth messages or "phone tag."

The same concept applies to e-mail. E-mail is one of the greatest disruptions to your workday. I advise managers to check and respond to their e-mail at one or two predetermined times of day—no exceptions. Looking to see if you have a new message twenty times a day is just not a productive use of time. Excessive e-mail use also violates the "like activity" law of calendaring. Reading and responding to messages will be quicker if you process multiple e-mails at one time.

I also advise against checking your e-mail first thing in the morning. Opening your inbox immediately upon arriving at work is a bad idea because it disrupts your concentration. Your brain is refreshed and relaxed in the morning, and diving headfirst into reactive work is not the best use of this optimum mental energy.

If you are uncomfortable allowing your e-mails to pile up throughout the day for fear of missing an urgent message, I have a simple solution for your dilemma: use the "out of office" instant response setting to alert clients and colleagues of your e-mail schedule:

Thank you for writing. I appreciate your message; however, my e-mail correspondence for the day is complete. I will read and respond to your e-mail at XX a.m. tomorrow. If this is an emergency or requires immediate response, you may leave a message with my assistant by calling (555) 555-5555.

Walk-in visits from team members also fall into the reactive work category. While restricting your "visiting hours" or requiring an appointment can be challenging for a project leader, having rules for when you are free to talk is critical. Whenever possible, encourage employees to schedule appointments with you for specific times, just as a client or other colleague would.

If you aren't comfortable with completely banishing drop-in visits from your team, establish specific times of the day where team members can and cannot just pop in. Make certain that your team understands that they are not to disrupt you for any nonemergency matter at certain times, such as when your office door is closed or before lunch.

Dealing with Other Unexpected Issues

Granted, not every unexpected task that comes your way is reactive in nature, nor can you batch all unexpected tasks into one neat block of time each day. For example, if your boss comes into your office and wants to chat, you have few alternatives but to oblige. Or, if a team member comes to you with a serious roadblock in the project that is preventing the group from moving forward, you may have to deal with the situation immediately. Similarly, if an important client or project stakeholder asks you to complete an urgent request, common sense dictates that you get whatever they've asked done as soon as possible.

Many people find their productivity completely stymied by interruptions, but the reality is that distractions are a part of life. The point is to rein in as much of the reactive work as possible and to rely on your Workflow Management Model to capture all of the other tasks and input that you would otherwise be handling. You have to be able to trust that your system is in place and that your to-dos are recorded so that you are prepared to tackle those tasks when you have the time available to do so.

You may not always be given perfect blocks of distraction-free time. There will always be unexpected situations to deal with, urgent calls that you have to take, and drop-in guests that cannot be ignored. The goal of the Workflow Management Model is to (1) ensure that nothing is overlooked when you're dealing with the unexpected, and (2) help you assess the best tasks to tackle when you're blessed with the gift of free time.

Choosing What to Do at Any Given Time

So what do you do when you aren't reacting to incoming calls and e-mails or handling an unexpected situation? Having your to-dos and calendars outlined gives you greater flexibility for determining the activities you perform in your day. In many cases, you will have some discretion to choose what item to complete first, which means that you may have to prioritize the activities you tackle in order of least to greatest importance.

If you are not working on a task already and are unsure what activity to tackle next, start by looking at your calendar. Do you have any commitments already scheduled for the day, or do you have complete discretion on how to best use your time? Once you've assessed your calendar to determine how much time you actually have available to work (or how much time you expect to have available), you can then review the items on your urgent to-do list.

If several items on your to-do list are of equal importance, you need a process to determine which activity to tackle first. When choosing the order of tasks to tackle, you must consider the following:

- *Ease of completion.* Some tasks can be completed anywhere, at any time, while others require specific tools, resources, or work environments. For instance, you can't send e-mails without an Internet connection, and you may not be able to conduct personal business on work time. Choose the tasks that your current environment is best suited for, and move on to the next criterion.

- *Time.* You must next consider two factors relating to time—how much time you have available and how much time the task will take. If there is a discrepancy between the two items—for example, you only have twenty minutes and a specific task will take at least an hour—you may need to select another activity to complete.

- *Mind-set.* Certain tasks require more mental effort and exertion than others. Some activities are best completed in the morning when your brain is fresh and alert, while others can be performed at any time with ease. When considering what to do next, evaluate your personal energy level and ability to function productively.

If you arrive at a meeting fifteen minutes early and are sitting in your car waiting to go into the building, it's important to determine the best, most productive use of that time. For example, you are unlikely to have the tools, time, or resources necessary to start working on your annual report to the company's CEO. Similarly, the time might not be right to make a personal call when you need to be mentally prepared for the upcoming meeting. However, if you need to make a quick call just to get information about a product price or to schedule an appointment, that might be the perfect task to tackle, given your current tools, state of mind, and fifteen-minute block of time.

Your to-do list should already be broken down into categories of like items (i.e., calls, appointments, things to do at the computer, and e-mails to send). If your to-do list items are grouped by the type of task, you should consider the above criteria for each group of activities. Working on similar items for a period of time (batching) will help you get more done faster. You'll be able to make seven calls quicker, for instance, than if between each call you took a break to respond to e-mails or write up your weekly reports.

Why "Time Management" Is a Misnomer

Everyone has heard of "time management." Almost every schoolteacher and CEO in America has lectured someone about the need to manage his or her time better, yet the very concept of time management is a misnomer. No one can actually "manage" his or her time, at least not in the traditional sense. You can make the most of the eight hours in your workday by getting in the flow and minimizing distractions, but even the most effective worker in the world cannot make those eight hours into nine (although he or she may accomplish well over eight hours of work). You can't control or manage the clock, but you can manage your actions in the time you are given to achieve more in less time.

You can achieve anything you set your mind to, so long as most days, most of the things you do move you closer to your goal. Action management is about learning how to choose the actions you tackle, as well as those actions you choose to delegate, delay, or drop altogether. Too many people become overwhelmed by all that they have to do. But you cannot just "do" some things. You can't do a project; you can only identify and complete actions that will move you closer to the finish line.

The purpose of action management is to identify and perform the activities that will move you closest to the finish line. Many people are not truly suffering from a lack of time; they are simply spending that time on the wrong actions. If you or your team is struggling to find the time to complete critical tasks, it's time to sit down with your calendar and evaluate your priorities. You can delegate, delete, reorganize, and prioritize your activities to ensure that you're spending your time on the actions that matter.

Many people rely on a calendar as the foundation of their organizational system; in fact, calendaring is a central component of the Workflow Management Model introduced in Chapter 12. Calendars are prevalent in the professional world for good reason. Visually depicting their daily schedule on a calendar helps busy leaders stay focused on their goals and minimizes unnecessary distractions. Without a functional calendar to maintain your schedule, focusing on your goals, completing your daily tasks, and planning for the future are nearly impossible. If you have not already begun using a calendar to coordinate your days, I strongly urge you to get one now. If you are already using a calendar, it's time to learn to calendar for maximum efficiency.

Calendaring 101

During leadership development workshops and other public speaking engagements, I strongly encourage participants to apply the organizational principles they use at work in their home lives as well. While maintaining a calendar for your home life may seem awkward at first, many people find that doing so actually improves their quality of life outside of the office. Truly embracing a calendared lifestyle will enable you to make the most out of each day and ensure that you have the time to do the things that are important to you. For example, if you want to spend more time with your children, spouse, or other loved ones, calendaring time for activities with your family will guarantee that your wish comes true. When you make calendaring a way of life, you will learn to make time for the activities that count and become more selective about how you choose to spend your time.

Getting Started

If you've never before used a calendar, follow these simple steps to start working on the first draft of your new scheduling tool. Even if you're already experienced in the art of calendaring, it would be beneficial to skim over the step-by-step instructions below for a quick refresher.

1. *Make a list of every single activity that you complete (or are supposed to complete) each day.* Do not leave out tasks such as driving, showering, eating, or going to the gym. Record every activity that you perform in a normal day, week, or month.

2. *Group the activities according to type.* Work-related activities should form one category, while activities you do for entertainment or pleasure would form another group. The number and type of categories you have on your calendar will depend on your unique needs.

3. *Consider each group of like tasks.* When is the best time to perform each of these groups of activities? When will you be able to perform the tasks at an optimum level of efficiency? Going to the gym or working out is a great example of a task that may require some trial and error for scheduling. Some people find that getting their exercise out of the way in the morning when they are alert and energized is a great way to start their day. Others might struggle to drag themselves out of bed an extra hour early in the morning and regularly skip workouts, no matter how great their intentions were when they scheduled the task. Find a time when you will be best prepared to complete each task, in terms of both time and energy.

4. *Look at each task individually and as a group.* How long will it take you to complete each activity? Now round up the time you've estimated (add anywhere from five to thirty minutes, depending on the task; most people have a tendency to underestimate the amount of time an item takes).

5. *Schedule the tasks on a calendar, maintaining groups of like items whenever possible.* For now, schedule your workday as a large chunk of time—do not specify exactly how you will spend each minute of your time at work. Later in this chapter, you'll learn special tips and tricks for getting the most out of your professional routine.

6. *Review your calendar and ensure that it reflects all of your obligations.* Have you scheduled time for dates with your significant other? Going to the gym? Going to the doctor? Visiting your parents or grandparents? Taking a vacation? Your daughter's recital? Be sure you've included every activity you need or want to complete.

A word of caution: The calendar you create initially may not prove to be the most effective schedule for your life. If you've never considered when the best time to perform certain tasks (for example, going to the gym) is for you, your schedule may need to be reworked to reflect your needs.

You may also find that you've scheduled more than you could ever possibly complete. After attempting to calendar their lives, most people find that they simply do not have enough hours in the day to perform every activity they want or need to do. Prioritizing activities in order of importance becomes absolutely essential. If your calendar is bursting at the seams and you lack the time to complete everything scheduled, you may have to do some soul searching. The following guidelines may help you downsize your calendar to make it more "doable."

- *Look at the activities for which you have scheduled the most time.* Are these tasks truly worthy of so many precious minutes (or hours)? You may not have any way around the ten-hour workday on your schedule, but consider other large chunks of time and evaluate whether they are really worth dedicating so much time.

- *Look for activities that could be delegated or removed from your calendar altogether.* If the task isn't absolutely crucial to your daily routine—or if there is someone else who could perform the task as well as you—delegate it or drop it from your schedule. Perhaps your kids could take on more of the housework. Your assistant might be able to handle your errands and other nonessential tasks. Aim to perform only the activities that truly require your attention.

Sticking to Your New Schedule

Once you feel that you've developed a schedule you can live with, you must begin adhering to your new routine. Writing down everything you want to do in a day is the easy part—actually completing all that you've set out to accomplish can be a bit trickier. There are many reasons why people fail at following the schedule they've outlined—we'll cover a few of the biggest problem behaviors in this section.

One of the most challenging aspects of developing a calendar is learning to make the most from your workday. Smart scheduling is essential to achieving maximum productivity at work, even when using the Workflow Management Model for getting things done.

Bringing Your Team on Board

Adhering to your calendar is more challenging if your friends, family, and coworkers do not respect your schedule. You must encourage those around you to honor your calendar and show regard for how you spend your time. Requiring others to respect your calendar will set a good example for your team and ensure that you are able to stick to your schedule.

If you have calendared specific times for appointments, phone calls, or e-mail, your team should abide by this schedule. Your assistant (or answering machine) will record messages and screen calls until you are prepared to process them. Team members can communicate with you through a similar system. Teach them to keep lists of questions, concerns, or comments that they wish to share with you, and then schedule a time for discussion. This practice will streamline your life, and make your team members more productive and efficient in how they manage their time. When team members are not permitted to disrupt you with every question, they will also become more independent and get better at finding the answers themselves.

Dealing with Distractions

Scheduling your life on a calendar is a great way to increase your productivity and efficiency; however, calendaring alone will not get rid of the distractions disrupting your day. Distractions prevent us from achieving the tasks we must complete or, at the very least, make these activities take longer than is really necessary. Interestingly enough, many people allow (dare I say even encourage) distractions to intrude upon their daily routine. Distractions—whether welcomed or simply accepted—limit your ability to get things done and waste precious time. If you want to make the most from your calendar and accomplish more each day, you must learn to eliminate or minimize the distractions in your life.

To get a clear understanding of how distractions are paralyzing your productivity, try this simple exercise. Record every task or activity you perform throughout one or two workdays in a journal. At the conclusion of the experiment, review the list of your activities.

- How many of the tasks on your list were high-priority professional actions or activities that moved you closer to a goal?
- How many of the activities on your list did nothing to further your progress on a project or toward a goal?

The activities in the second category are most likely distractions that prevented you from doing the work that truly matters.

Distractions are dangerous for a number of reasons:

- They break our focus.
- They waste time that could be spent on more important or desirable activities.
- They disrupt our schedule and limit what we can accomplish in a day.
- They can become part of your routine, creating a bad habit.

As we discussed in previous chapters, getting in the zone and focusing on tasks with the maximum level of concentration will allow you to be a more productive leader. When you are working on a calendared item, you should be concentrating on that activity and ignoring any other (nonemergency) elements of your work environment. Answering the phone, checking your e-mail, surfing the Web, or getting drawn into small talk will all prevent you from working effectively.

Tips for Minimizing Distractions

Minimize Nonproductive Conversation

Other people can be one of the biggest distractions in the workplace. Some employees will always view their job as a place for socializing, rather than a place for working. If pointless chatter is constantly disrupting your day, you may need to learn to minimize small talk or superficial conversation without being rude or offensive.

One of the best ways to make conversations shorter and more productive is to simply ask for what you want. Getting to the point of a conversation is not impolite; in fact, many people will actually appreciate your brevity. Directing a conversation toward its conclusion is quite simple, especially if you start from the beginning of the discussion. For example, if you need to ask a team member who tends to be long-winded a quick question, try saying something like "Hello John, how are you? I know you're a busy guy, so I'll just get right to the point. What do you want to do about ..." You can also apply the same technique to phone calls to minimize unnecessary chitchat.

If someone calls or otherwise contacts you, you can use a similar approach. "Hey Megan, what can I do for you today?"

Stay off the Internet

The Internet has become one of the biggest distracters of modern life. It's easy to wander aimlessly through the Web, browsing site after site with no real purpose. To prevent the Internet from draining your time, avoid logging on as much as possible. If you do not have to use the Web for a specific, work-related purpose, just stay offline. You can also dedicate specific times of the day to getting on the Internet, such as over your lunch hour or as a reward at the very end of the day—just remember to abide by your schedule, and do not browse beyond your allotted time.

Eliminating distractions from your life is not an easy task; many distractions have become regular parts of our routine. If you struggle to stick to your new rules and your new schedule, focus on the benefits of the new system. Abiding by a calendar will allow you to accomplish more in less time and help you complete projects faster. You're not going to lead your team to new heights surfing the Web or responding to e-mails all day—you need to focus on the most critical activities demanding your attention!

Overcoming the Procrastination Problem

For some people, distractions are a symptom of a larger problem: procrastination. Procrastination is a huge problem for our society; in fact, some people battle the tendency to procrastinate for their whole lives. Why do people procrastinate so much? And how can leaders learn to limit this unproductive behavior, in both themselves and their followers?

Procrastination generally has origins in certain personalities, such as perfectionists, Type A personalities, adrenaline junkies, and highly intelligent individuals. The tendency to procrastinate can also be a result of other factors, including lack of self-control, an over- or underestimation of how long it will take to complete a task, fear, or a sense that the time is not right to do the job.

The Perfect Procrastinator

Perfectionists and Type A personalities frequently struggle with procrastination in the sense that they never finish a task because these people view their work as less than perfect and thus never finished. Perfection-minded individuals may also delay starting a project because they are constantly fine-tuning their plan to get started. You might say this is procrastination as a result of "getting ready to get ready."

The Adrenaline Seeker

Other people procrastinate because they believe they need the adrenaline rush that comes with working under pressure. This is a common feeling among students, who later carry this unfortunate belief with them into the workplace. Some people believe they thrive when working against an intense deadline; in most cases, they simply haven't tried to work any other way because they always procrastinate. Adrenaline cannot lead to a better result than a carefully planned and executed project.

The Genius

As odd as it may sound, procrastination can be the direct result of a high level of intelligence. Smart people actually tend to procrastinate more because they are all too aware of the implications of the project. They've already identified every thing that might go wrong in the project—every place for delays, every impossible task, and every roadblock that they will encounter during the undertaking of the task.

When a highly intelligent person sees a project, he or she understands its depth and the implications of the work. Whereas a person of average intelligence might be able to plunge in and start moving forward, geniuses might be stopped in their tracks as they consider all that is at stake.

Traits That Lead to Procrastination

Some procrastination traits are not characteristic of a certain personality type, but rather bad habits. For example, many people simply lack the self-control or self-discipline to focus on a task. Therefore, they continue to postpone the activity until there is absolutely no room left for delays.

An inaccurate assessment of the time it will take to complete a task can be another contributing factor to procrastination. If someone believes that a task will take a long time, they may delay it to wait for the "right" time. On the other hand, if that person believes an activity will take no time at all, they may push it off to the last minute. In either case, this poor estimation of time leads to postponing the work.

Finally, fear can be a major contributor to the procrastination problem. The procrastinator may be afraid of the unknown, afraid of failure, or even afraid of success. While this might seem like a ridiculous notion, fear is one of the biggest motivators of human action (or lack thereof).

Chapter 14 Review

The final key to getting things done arises from the task level of the team triangle. The activities performed by your team determine their level of success. Tasks are the engine that drives a project forward or runs it off the rails. The project leader must lead by example and help the team make good decisions regarding what activities to work on when, and what to delegate or drop altogether.

Before embarking on any task, the team member who will be performing the activity must understand the what, why, and how of the task. What is the task, why does the task need to be performed, and how is the task supposed to be completed? The team member should understand quality expectations, guidelines for performance, and any ethical issues to consider. The team should also understand where each activity falls in terms of priority—is this the most important task to be working on at this time?

Batching versus Multitasking

One of the biggest things a team can do to improve their efficiency is adopt the concept of batching. Batching involves performing like activities in sequence, rather than jumping from one type of task to another with no rhyme or reason. Team members can batch groups of like activities by breaking down tasks into their steps and grouping like activities. They can also batch by scheduling and calendaring specific types of activities to be performed at a certain time.

In no way should batching be mistaken for multitasking. While many people may pride themselves on their multitasking skills, the reality is that multitasking is a myth. When someone attempts to multitask, they are really just switching back and forth from one task to another. Multitasking actually results in decreased productivity, wasted time, and reduced focus. Multitasking should be discouraged at all costs.

Calendaring

Learning to honor the calendar is one of the most important things a project leader can do. Project leaders must learn to calendar if they want to operate at optimum efficiency. Creating a calendar that batches like activities and restrains reactive activities can help project leaders get more done in less time. Rather than wasting time considering what to do or switching back and forth between tasks, a calendar enables leaders to focus on the right tasks at the right time. A calendar also helps leaders prepare for the unexpected and ensures that nothing is overlooked.

While there is no way to create more hours in a day, project leaders can make the most of the time they have by working on the right tasks at the right time, eliminating distractions, and learning to overcome problems with procrastination.

Conclusions

> You can accomplish anything in life, provided you do not mind who gets the credit.
>
> **—President Harry S. Truman**

The main premise of this book has been that companies need more project leaders, and fewer to no project managers. Project leaders lead projects in much the same way that successful entrepreneurs lead companies.

Successful team leaders start with vision, passion, and working through their teams. They then use appropriate tools for assembling their team, assessing their potential, and mediating conflict. The role of a team leader is that of a mediator, guide, and servant (as opposed to a boss, manager, or disciplinarian). The team leader understands that at the end of the day, it all comes down to people. The true test of leadership is, first, whether anyone follows; and, second, whether those who follow are better for having done so.

The challenges of a project leader are numerous: they must deal with deadlines, personality conflicts, the expectations of superiors, internal conflicts, and the ever-inevitable change. Still, these challenges can be met with the appropriate tools if the leader has started with a secure foundation.

What I hope has emerged from this book is the idea that leadership is not something granted or a special gift that only a chosen few are born with. Leadership is built, developed, nurtured, or, better yet, grown. Some of the greatest leaders of our time come from humble beginnings—and many exhibited no signs of excellence in their early years. The greatest leaders share one quality above all others: a commitment to improving, to raising their game, and to becoming a better person. Project leaders work hard at leading others way before they are actually given a project to tackle (or a fancy title on their office door, for that matter). And, like a delicate flower that grows in a desert, leadership must be nurtured and appreciated every day.

Appendix

The aim of this appendix is to give you some concrete, actionable tools for use in leading your team. Some of these exercises will be helpful when you first assemble your team, some will be helpful after a project is underway, and some are meant to be small, private exercises that can be done at any time. Consider incorporating some of the team exercises into your next retreat or meeting, and your team will evolve before your eyes.

Team Exercises

Perhaps the best weapons in your leadership arsenal are group or "team-building" exercises. These exercises are designed to get team members involved while giving you powerful information you can utilize as you serve and lead your team.

Individuals will react in different ways to team exercises, but the overwhelming reaction of most employees is typically skepticism. This skepticism is a stark contrast to the attitude of the team leader. Most leaders will approach an exercise with a "Hey everybody, look what I got!" sort of mentality—in other words, they will be enthusiastic and hopeful, especially if the exercise has sparked some new ideas in their minds.

Employees typically do not come to the table with the same enthusiasm and inspiration. Indeed, many employees will be soured on such "team-building exercises" due to bad experiences with them in the past.

Some caution is needed, then, in introducing new exercises to your team. Try to keep the following in mind when considering the appropriate time and place to introduce an exercise:

- The best time of day to try a new exercise is in the morning, before the workday has actually begun (and after everyone has had some coffee!). If you try an exercise at the end of the day, people might be tired or just ready to go home. If you run an exercise midday, you risk interrupting work flow and have the added hassle of capturing and maintaining your employees' attention. Start your activity at the beginning of the day before your team is distracted, and while you have their undivided attention.

- You don't need to explain your overarching reason for conducting the exercise. This risks sounding overly vague, or too "touchy-feely."

But you should say something about the purpose or goal of the activity. Give your team just enough information to structure the activity, and then stop.

- People are usually hesitant to join into an activity unless they see others start first. Before an activity, speak privately with a member of your team who you trust. Relay to him or her a brief introduction to the activity you are thinking of trying, and ask for help. Get him or her to commit to "jumping in" once you've explained the exercise. Others will follow his or her lead.

- If you have a trusted and capable team member, consider having that person lead the activity instead of you. This shows employees that you trust them and want to see them in a leadership role—talk about empowerment! But delegating the activity to another also makes the activity look less like your pet project, and more like a team-building exercise. Finally, delegating allows you to participate fully without worrying about how the activity is run, but still leaves you with the power to step in and run things if they don't go smoothly.

- Every activity you try should be something you do yourself first (as far as possible). You need to set the example for the other team members by showing a willingness to participate and demonstrate familiarity with the activity. Doing a dry run can also reveal potential questions, problems, or issues.

- Don't try too many activities at once (or in a row). Try one exercise, take some time to evaluate how things went, and then hold off on other team-building exercises for a couple of days. Then reevaluate the activity once some time has elapsed and you have a fresh perspective. Only then should you move on to the next activity.

Most importantly, you will need to have your own attitude tuned to the activity in the following ways:

- *Being committed to the exercises.* See activities through no matter what, and make it clear—to yourself and the team—that these will go on for some time.

- *Being real.* Don't dress up the activities as special, don't use buzzwords or business-speak, and don't break out the expensive multimedia equipment. Treat these activities just as you would any other project or task your team has to do.

- *Being ready to follow up.* Are there results that need to be communicated? Action plans or promises that need to be followed up on? Reports to write or share? Integrating an activity must be done at both ends—from the time you first present an activity to the point where you can see tangible results.

In the pages that follow, you'll find exercises to help build your team's IDEAS:

Identifying attributes
Debating essential issues
Embracing accountability
Achieving commitment
Setting and maintaining standards

Identifying Attributes

The 360 Review

Purpose: To give a well-rounded assessment of a team member's potential, especially leadership potential.

Time required: Variable.

Procedure: Acquire or devise a 360-degree review for a team member (links to professional 360-degree assessments are given below). Give this review to (1) one "superior" or manager in the organization, (2) two of the team member's peers, and (3) two of the team member's subordinates. After each has completed the review, compile the results and go over them with the team member.
  Premade and validated reviews can be found at the following sites:

- www.decwise.com/sample-360-degree-feedback-survey.html
- www.completesurvey.com
- www.reactive360.com

Variation: Once you have assessed several team members, do a 360 review for yourself. Have a neutral party compile the results and share them with you. Then share the key findings with your team. Review the section in this appendix on "Opening Your Kimono" for further suggestions on this variation.

Personal Histories

Purpose: To build trust in a nonthreatening way and help team members recognize each other's histories and grounds for decision making.

Time required: About ten minutes per team member.

Procedure: Go around the table and have each team member share (1) where they grew up, (2) how many siblings they have, and (3) a moment from their childhood that was challenging, scary (in retrospect), or enlightening. The goal is for everyone to share something personal but not threatening, that is, something without "TMI" (too much information).

Variation: If you feel bold, one great variation on this activity is the "Two Truths and a Lie" game. Have each team member share three pieces of information about him or herself—two facts that are completely true, and one made-up "fact" that is completely false. Have the other team members then try to guess which statement is the false one. This encourages team members to share a lot of information and rewards other employees for listening.

Behavioral Profiles

Purpose: To give team members insight into their own behavioral profiles and those of other team members.

Time required: Variable.

Procedure: Before meeting with the team, have each member complete the Golden Personality Type Profiler or a similar behavioral assessment tool (make sure that you use a professional tool and not just one of the slew of "free" assessments on the Internet!). When your team meets, explain the test and how the results should be read. Give the results of the test to your employees and allow them to ask questions. Remember, scores should be confidential, but you can encourage employees to compare results with each other of their own accord.

Variation: After results are distributed, ask your team members which results they would have preferred, if not the ones they actually got. Ask your team members what this says about their self-perception, their goals, and what they admire in others. More often than not, each personality type is envied by some other!

For several excellent providers of this tool, please see our companion website, www.ManagementToLeadership.com.

Debating Essential Issues

Conflict Resolution Exercise

Purpose: To give team members practice in appropriate conflict resolution.

Time required: About one hour.

Procedure: Choose an issue that the team has faced recently and that was particularly difficult to solve. At this stage, you should choose a problem that has been solved, put on hold, or otherwise dissolved—choosing an ongoing problem is risky and will threaten to derail the exercise.

Have each member review prior discussion of the issue and identify particular obstacles to resolution. Try to categorize these obstacles as far as possible. Particular obstacles can fall under categories like the following:

- *Information obstacles*: Lack of facts, diverse perspectives, untimely information
- *Motivation obstacles*: Poor corporate culture, poor leadership, uninspiring activities, lack of clear goals, lack of trust
- *Organization obstacles*: Poor planning, vague steps or activities, unclear delegation of responsibilities, lack of metrics (or poor metrics)
- *Activity obstacles*: Physical constraints, unrealistic deadlines, supply problems
- *Personal obstacles*: Personality conflicts, differing expectations (or unreal expectations), lack of appropriate skills, lack of commitment

Once team members have listed and categorized specific obstacles, have them compare answers and discuss what impact these obstacles have on team dynamics and problem solving. Then brainstorm ways of overcoming the most common obstacles in the future.

Variation: If using a current problem in your organization might leave things too "charged," ask team members to pick another obstacle from their lives, or simply use a obstacle from the news, a history book, or the like.

Conflict Continuum Exercise

Purpose: To give explicit feedback to team members about their potential for conflict and possible areas of improvement.

Time required: Thirty minutes.

Procedure: Have team members write their names at the top of a piece of paper. Then have them divide the paper into three columns. At the top of the first column, write "Frequency"; at the top of the second, write "Importance"; and at the top of the third, write "Intensity."

Then have each team member pass their sheet of paper to the person to their left, whom we will call the "rater." The rater must now place an "X" in the column for where that team member falls in each category with regard to his or her potential for conflict.

If the named team member creates conflict often, the rater should put an X near the top of the "Frequency" column. If he or she rarely starts conflicts, the X should be near the bottom.

If the named team member engages in conflict about truly important things, an X goes near the top of the "Importance" column. If the named team member tends to engage in conflict about petty things, have the rater place the X at the bottom.

Likewise, if the named team member engages in conflict in ways that are intense, angry, dramatic, or otherwise emotional or personal, have the rater put an X at the top of the Intensity column. If not, have him or her put the X near the bottom.

Once each rater is done, pass the papers to the left again and continue these steps until everyone has had a chance to rate everyone else. At the end, each paper should have a series of Xs in each column.

Chances are that these Xs will tend to cluster together. Have team members review their own charts, and try to see if there are any patterns. Then have them summarize their results for the team and discuss the implications of the team's conflict style as a whole.

For a template for this exercise, please consult our companion website, www.ManagementToLeadership.com.

Thomas–Kilmann Model

Purpose: To help team members understand their habits of conflict and collaboration, and to encourage them to approach conflict consciously (instead of relying on their natural, and possibly inappropriate, reactions).

Time required: Variable.

Procedure: The Thomas–Kilmann model identifies five different kinds of personalities with regard to conflict and how conflict is perceived in terms of tasks and relationships.

The model and assessment can be found at www.cpp.com.

Embracing Accountability

Team Effectiveness Exercise

Purpose: To provide team members with focused, explicit feedback and suggestions for improvement.

Time required: About thirty minutes, depending on the size of the team.

Procedure: Have each team member write down the following for each other member of the team:

1. A specific behavior, set of behaviors, or behavioral tendency that contributes positively to the team and project (i.e., the team member's strength).
2. A specific behavior, set of behaviors, or behavioral tendency that causes conflict or otherwise proves problematic to the team (i.e., the team member's weakness).

Collect these answers and give each team member his or her feedback. Beginning with the team leader, have each team member read aloud his or her strengths first. After each member has read the list of strengths, ask them to comment about them (but encourage them to do so in a positive way). Were there any surprises? Anything that was confusing or contradictory? How did that list of strengths make the team member feel?

Next, ask each team member to read his or her weaknesses aloud as well, starting with the team leader. At the end, have the team member summarize those weaknesses (try to get this down to one or two key points to remember). Then ask everyone to come up with a few ideas to correct or compensate for those weaknesses, and have them e-mail you the results.

Variation: Try to identify pairings of team members whose strengths and weaknesses complement each other. For example, if one team member is viewed by all the others as being disorganized, find a team member that is known for his or her organization. Have those team members work closer together, or perhaps enter into a kind of "equal mentorship" arrangement.

Team Scoreboard

Purpose: To provide an explicit and tangible measure of the team's success.

Time required: One to two hours.

Procedure: First, establish two to four general activities that are critical to the team's ability to meet its stated goals and objectives. These need not be concrete; activities can include things like "Establish consistent pricing," "Develop a low-cost marketing plan," or "Expand into new sales territories."

Once you have a handful of critical activities, discuss metrics that can be used to measure each activity. Metrics should be concrete things that can be measured, and should indicate whether or not the right steps are being taken before results are in, and not after.

For example, if one of your activities is "Expand into new sales territories," do not set the number of sales in a new territory as a metric. Instead, measure the number of sales calls in new territories, or the number of leads gathered, or response rates to new advertising.

Another example: suppose your critical activity is increasing market awareness. Do not measure inside sales calls or customer awareness. Instead, measure progress on concrete marketing activities: developing the brand, placing advertising, going to trade shows, and the like.

Keep a running checklist of these metrics, and update them on a daily basis.

Variation: If you are inclined toward numbers, developing a brief report based on your metrics can be fun and informative—not to mention that such a report is useful for explaining results to higher-up management.

Achieving Commitment

Clarification of Organizational Principles

Purpose: To generate a list of core organizational principles that clarify the common goals and perceptions of your team.

Time required: Several hours, depending on your team and the issues that are raised.

Procedure: Have your team discuss the following issues and come to a common consensus on the appropriate aims and goals:

1. Core purpose
2. Core values
3. Vision for the future
4. Distinctiveness of the organization and competitive advantage
5. Business model or strategy
6. Overarching goals
7. Long-term activities that meet these goals
8. Daily practices and steps that help accomplish long-term activities
9. Metrics for measuring success
10. Roles and responsibilities

Clarification of Team Principles

Purpose: To generate a list of core team principles that clarify and direct the ways in which members of your team deal with you, with each other, and with conflict.

Time required: Several hours, depending on your team and issues raised.

Procedure: Have your team discuss the following issues and come to a common consensus on the appropriate procedures and goals:

1. *The length and structure of meetings.* What roles do people play? Who does what, and how is the agenda set?
2. *Etiquette for meetings.* When should questions be asked? When is laptop use OK? Are interruptions encouraged?
3. *Time and flexibility.* How important is being on time? How rigid or flexible are schedules? Deadlines? How available are team members expected to be during nonwork hours?
4. *Resources.* How are resources distributed, and by whom? What is the procedure for using common resources? To what degree can team members engage one another's staff and resources?
5. *Communication.* What are the preferred methods of communication? How should sensitive or confidential communication be handled? What kind of turnaround time is considered polite or timely for these communication methods?
6. *Grievances.* How should complaints and problems be handled? To whom should grievances be reported? What processes should be in place to handle them quickly and confidentially?
7. *Etc.* Try to think of any other issues, policies, or activities that team members might engage in on a daily or almost-daily basis.

Variation: Make the "Clarification of Team Principles" exercise a regular part of your meeting schedule. Since there are many issues to cover, the best strategy is to take one issue or set of issues and use the first twenty minutes or so of regularly scheduled team meetings to address them.

Setting and Maintaining High Standards

Resumé Exercise: The Perfect Team Member Is You

Purpose: To achieve consensus about team excellence, and to commit to higher standards.

Time required: About an hour.

Procedure: Have your team pretend that they are going to hire a new team member, and have them generate a fictional resumé for an ideal employee or team member. They may generate the resumé from scratch, or else have each individual create a resumé and then have the team compile the results.

If there are items on the ideal resumé that match the qualifications of actual team members, point this out. Many of the qualifications of an ideal candidate likely exist on your team already!

Now have team members think of three (or so) ways in which they can make their own personal resumé match that of the ideal. Should a given team member take a class or seminar in a particular area? Does he or she simply need more "frontline" experience? Is there a small subproject that a member can coordinate in order to round out his or her list of accomplishments? Help your team members develop a plan of action for becoming the ideal that they envisioned together.

The "Tower of Babel"

Purpose: To exercise leadership and role-playing skills, and to identify team dynamics.

Time required: One to three hours, depending on the team.

NOTE: This activity requires quite a bit of time and space, as well as some preparation on your part. Consider doing this as part of an off-site workshop.

Procedure: Before gathering your team, you will need to make up some index cards with a number of "roles." Each role should define a way in which a team member is to communicate, and should be fairly difficult. Some ideas would be the following:

- "You can only communicate in a foreign language."
- "When you speak, you must be negative and/or sarcastic."
- "You can speak normally, but cannot touch anything."
- "You must play devil's advocate whenever possible and challenge what is going on."

On one card, the "leader" card, put something along the lines of the following:

You are the leader. You must communicate with your group without talking or writing anything down. Your task is to have the team build the tallest tower they can that will support the weight of a book when finished.

Everyone has a different role, and cannot speak or write except as specified on their role card. Team members will have to find a way to communicate with these different dynamics going on. Good luck!

Once the cards have been made, gather your team and put some "materials" in the center of your work area—good materials would be straws, index cards, Scotch Tape, and the like. Then distribute the role cards randomly to your team. Each team member must read his or her card and stick to the role given there. Team members cannot share their role (or card) with any other member—they know their own roles, but no one else's.

Then tell your team to begin, without any further instructions. The job of the leader will be to organize the team and try to involve everyone in the project. Take note of how the leader does this, as well as how the other team members interact.

When the activity is completed, talk about what went on during the process. Stress the importance of inclusion. Talk about some of the problems that they had with communication—and the perceived differences in what, exactly, the problems were.

Leadership Exercise: Role-Playing the Medieval Court

Chapter 2 mentioned the benefits of role-playing particular roles that you might be unfamiliar or uncomfortable with. The following exercise is designed to give both structure to this idea and inspiration to you as you lead your team.

The Idea

In medieval times, kings and feudal lords were chosen because of their bloodline (or rose to power through brute force). Consequently, the person sitting on the throne was not always the most qualified or most experienced leader. Still, the king was ruler of the land; and, like a project manager, a king had to fill many roles deftly and simultaneously.

Most kings overcame their lack of experience or qualification by "outsourcing" many of their roles to trusted officers that were more qualified. These officers comprised the king's court. Each member of the court had a specific role, source of power, and outlook on the problems and resources of the kingdom.

Over time, the members of a king's court became symbolic of certain roles and duties generally. Although you will probably not have the benefit of an entire court at your beck and call, you can learn about the various roles in the court and learn to "switch hats" as needed to serve the members of your team.

Directions

Read through each project management role, and then select the part that you feel least comfortable with. Beginning with the most challenging role first will provide you with extra time to grow accustomed to the role. Remember: part of project management is subjugating your needs for the good of the team. You may not be comfortable playing the role of the "Black Knight," but your team needs you to fulfill all of your roles—not just the ones that are easy for you.

Focus on assuming your selected role until the character seems natural to you. At that point, you may select another role to practice. Continue until you've familiarized yourself with every position in the medieval court.

The Zealot

On occasion, a project manager must be a zealot—an individual so committed to the process that nothing can deter him or her. In the beginning of a project, the project manager must exemplify the overarching attitude and actions of the team's task. He or she must doggedly follow the agreed-upon processes in order to emphasize their importance to the rest of the team. By continually reinforcing the same structure over and over, the zealot leads by example, giving team members a clear and consistent understanding of what needs to be done.

This dogged adherence to the established policies and procedures allows the project manager to hold his or her team members accountable. If a project manager agrees with the team that they are going to follow one process and then proceeds to follow a different set of rules, the team will assume they can do the same.

The project manager can only hold a team accountable to the rules by clearly defining his or her expectations for procedure, and then following those processes with incredible zeal and fervor.

The Bazaar Owner

In medieval times, there were individuals who sold wares in a central area or on the streets of a walled city. These bazaar owners made their living by not only understanding what residents needed on a day-to-day basis, but also ensuring that the merchandise got into the city and into the hands of the residents.

The only thing this owner had was a small stall in the market—and an uncanny ability to read the needs and requirements of a population. Today, we call our bazaar owners "entrepreneurs." These individuals look for opportunities to provide the best services and products possible to a population hungry for their wares.

A project manager must be an entrepreneur within his or her own project. The project manager is provided with people, a budget, and a timeline to meet the objectives. Project managers must be willing to negotiate for needed resources, understand the political and environmental landscapes, and provide his or her customers and team with the necessary motivation to achieve that goal. While project managers are often thought of as individuals who follow lockstep in a process, in actuality, the role of a project manager must be one of an entrepreneur—working in a dynamic environment to achieve results in a moral and ethical way.

The Scout

One of the keys to Britain's expansion was the empire's keen use of explorers and scouts to explore new territory. Although you won't necessarily be exploring new places as a project manager, you will need to learn how to navigate new information and new opportunities.

As a project manager, you will sometimes need to be an information gatherer, not just receiver. Your job is to scout ahead and observe what comes next. What is the next step in the process? What are the potential dangers and pitfalls? Though the scout faces the risk of the unknown (including the competition!), the benefits are well worth the risk.

A scout must also bring information back to the organization and participate in altering any plans. The information you bring back may change the course of the project, reveal the assumptions that you've made, or activate a risk plan. Although the scout is often the least visible role within an organization, he or she is vital for keeping the project on the path—and keeping it from falling off cliffs.

The Oracle

In ancient Greece, kings would travel, sometimes great distances, to visit the oracle at the temple of Apollo—usually before going to war or undertaking a serious issue. The oracle would "contemplate" the king's decision and after several days emit some strange sounds. These sounds would be interpreted by the priests and the person would be given a real suggestion that pointed them in the right direction.

Although you do not have the luxury of being cryptic, you are still looked upon as the Oracle: the individual who needs to see past the complexities and worries of an issue and discern the underlying pattern. This is a difficult task, but as a project manager you should be able to "see the future" and identify problems, trends, and patterns. While most teammates work on part of the project, you must be able to see and communicate the whole.

Consider the story of the foreman charged with cutting a road through the jungle. He works with people up front hacking away at the brush, and people who are making sure the road is straight, and still more people moving the waste and laying the asphalt. So where is the foreman? He's the guy in the tree, looking ahead and making sure that the road is going where it should!

The Knight Protector

In medieval times, the knights of the realm were sworn to protect and defend the kingdom at all costs. This entailed not only physically protecting the property and people of the land, but also defending the honor, dignity, and perception of the kingdom.

A project manager is not only responsible for getting the project done on time, on budget, and within the defined scope—but also responsible for maintaining the team's morale. Therefore, the project manager needs to be the knight protector, or champion, of his or her project. The project manager's actions can help or hinder the team's success. If project managers do not follow a standard project management process, they can hurt the project, but they can just as easily hurt it more if they don't understand the requirements of a project champion.

A team looks to the project champion to get clues on how they feel about the project's success. Once an individual is assigned a project, he or she must get behind it as quickly as possible. This means understanding clearly how this project works within the organization and with all of the other projects that are happening.

The project champion must keep a positive attitude and display that attitude at all times. A project manager is on stage every time he or she walks into the organization. Having a positive demeanor is critical.

The project manager must be the individual who not only understands why this project is important but also continuously communicates this to every member of the team—and every member of the organization. In so doing, they are not only making the importance of the project clear, but also reinforcing the reasoning for the project's existence. This reasoning is necessary, especially for long-term projects, since people have a tendency to forget the end goal and focus on the little items needing to be done.

The project manager must also be the champion for the project outside, dealing with the senior executives and continuously reminding them of why exactly this project is important and how they can be part of the solution.

The Scribe

During medieval times, the Scribe was probably the most literate individual within the entire court (with the possible exception of the priests). Though nobles and the king could read (albeit very slowly), they were usually unaware of all of the requirements of courtly language in the written form. Therefore the scribes were expected to write and read information proficiently.

Most of the scribes came from the priest class and were steeped in the traditions of the Church. Because they were proficient with several languages, including the language of state (Latin), the scribes served as a bridge between differing countries and areas. For example, when the Queen of England went to speak to a rebel queen in Scotland, the only language they had in common was the Latin that the priests had taught them.

In many ways, a project manager must be the bridge between a project team and all of the other organizations. He or she is also responsible for being the repository of all of the information for the project, knowing that the closure of the project becomes one of the most powerful things that a project manager can do. Finally, a project manager comes from a well-documented tradition of the project management process and helps other individuals translate that process into success.

We as project managers don't like to think of ourselves as scribes, but it was a substantially vital role within the court during the Middle Ages. It is our job as project managers to capture and deliver information in a coherent and accessible format, exactly like the scribes of the Middle Ages.

Father Confessor

The Father Confessor is an individual sworn to secrecy, but who knows all of the sins of the organization. If a project manager is to be successful, he or she must understand the true issues involved in the project. And sometimes these issues are not ones that should be brought into the public sphere. Therefore, a project manager must have a reputation of being able to keep information confidential when needed.

Garnering a reputation for confidentiality will serve you well beyond any one project: all organizations have public and private projects, and a project manager that shows his or her ability to keep information quiet will certainly be trusted with the more sensitive projects within the organization.

Another reason to keep information quiet is that project managers may have sensitive information that could damage the organization either financially (such as through stock movements) or through reputation within their industry. Because of their role in the organization, project managers are exposed to a great deal of sensitive information that must be kept quiet.

These project managers need to be able to capture the information to make the project successful, but not speak about this information. A project manager must truly become a confessor of all the bad things and only a promoter of the good.

The Flag Bearer

During the Middle Ages, illiteracy was widespread and most individuals, including the nobility, could barely read or write. To combat this problem on the battlefield, the nobility would have standard bearers carry the flags or insignia of each group into battle. With a single symbol to rally behind, the entire regiment would know where and who they were.

In medieval Japan, this was taken to an extreme. Each samurai (knight) would carry his family crest, either attached to his back or by a standard bearer. Most of the time, knights were not killed but rather held for ransom, and so being identified as a knight increased one's chances of survival—and of being reunited with family.

Project managers do not have to worry about kidnapping or ransom, but they do need to deal with carrying the appropriate message to the troops. The project manager must communicate the vision and purpose of the team, both to the team and to outsiders. The project manager must also learn the roles of other teams in the organization, and understand the role of his or her team within this bigger picture.

The Artisan

In medieval times, princes and kings often sponsored and encouraged well-known artisans. In return, the artisans gave them lasting tributes to their power and prestige.

In many ways, companies still do this by retaining certain project managers who have the unique ability to manage specific types of projects. Certain project managers, through their personality or their experience, are head and shoulders above other project managers on certain tasks. Healthy organizations recognize the need to encourage these "artisans," so they can perfect their craft and add to the overall value of the company.

An organization that has several individuals who can deliver on specific types of projects continuously and effectively has a real treasure in its midst. The challenge always is to keep these individuals interested and motivated—not only to do the best within their relative area of expertise, but also to mentor new individuals and keep successful practices alive.

A wise corporation will give these artisans special projects that they're good at, as well as other projects that they can work on, in order to expand their careers and grow at the same time. Artisans within the project manager process are hard to find, but once they are found, these individuals can be very valuable to both the team and the organization.

The Tactician

Historically, the role of the tactician was to assess the team's available resources and determine the best way to deploy these materials for optimal efficiency and effectiveness. Most tacticians were employed during wartime so that rulers could utilize their knowledge and skill to win a battle. In many ways, tacticians are still in demand today.

Project managers must be tacticians—skillfully using their knowledge of what the company needs, the requirements of the project, and the finite resources they have at their disposal to achieve the desired results. Project managers are shaping their organizations' destinies, but that is not an easy task without information. If the project manager is not aware of the needs of the organization, developing projects that support that process becomes a very challenging task indeed. The project manager must be able to understand the complete field and how he or she will get the organization to the next step or how the project fits in with the overall strategic plan.

When project managers have the big picture and the plan, they are also able to see opportunities that may be leveraged during a project. If not, these opportunities may be squandered and organizations can miss out on the leveraging power of a project manager.

The Historian

Of all the roles a project manager will fulfill, the position of historian will have the most benefit to the corporation in the long term but, regretfully, is the role that is emphasized least. During the battles of the Middle Ages and Napoleon's reign, the battlefield commander would keep a historian with him so that if he won the battle, the commander was able to dictate to the historian the definitive record of how the battle was fought.

Organizations need historians as well. The difference between the historians of olden times and the historians of today is that project managers must be able to record the history whether the project succeeds or not. In fact, recording the execution of a project becomes even more important when a project does not succeed, in the event that a similar project is assigned in the future.

Corporations would do well to keep records of not only how the project was intended to be implemented, but also how the project was actually executed once the task got underway.

The project manager as historian should identify not only what aspects of the project went well, but also what aspects of the project went poorly. Future project managers will be able to learn from the previous teams' mistakes, understand the risks, and execute the current project more efficiently. If an organization truly wishes to improve, then the project managers must act as historians by examining each past project to understand three things: what went right, what went wrong, and what could go better next time.

The Bard

In the medieval court, the legends and history of a civilization were told by the bard. The bard kept the institutional memory of the culture alive by sharing important messages time and time again through pantomime, story, song, and verse. The bard reminded his civilization to remember the more noble things in life and the virtues that the civilization's founders upheld with the highest esteem.

A project manager must be the bard for the rest of the company. Project managers are given a great deal of power and are changing the way people do business. Therefore, project managers must not only understand the culture but also continuously reinforce that culture through their actions, words, and deeds. Project managers must lead by example and help their team members capture that corporate culture and infuse its essence into every project.

As a bard, you should always be looking for positive actions that reinforce the hallmarks of your organization's culture. Once you see one of those actions, immediately encourage the team member and offer specific, positive praise about his or her actions. Then, share the story with the rest of the team. This reinforces that culture with the individual and, more importantly, helps other team members understand what is expected of them. Large organizations tell stories to reinforce corporate culture. A project manager can do more teaching by telling stories than by giving long lectures. What stories do you have to tell?

The Astronomer

In some ways, a project manager is very similar to a medieval astronomer in that he or she is asked to look into the future, predict potential issues and problems, and prepare for these scenarios before they arise.

Throughout history, individuals have used lessons-learned documents to predict the future and compare it to the present. Ancient astronomers used all of the documentation that they could possibly find to try to interpret the heavens and draw conclusions about what was happening on Earth.

An astronomer would look into the sky and use notes from previous observations in an attempt to determine the path of the planets. A good project manager will do the same by using previous observations to determine the trajectory of the project. The best documentation for project managers is the records they created in their role as historian. Astronomers understood that historical information was very valuable. Don't forget to look to the past to find answers for the future.

The Heir Apparent

The heir apparent was the individual who had been nominated to take over the throne once the ruling monarch has died—and the heir apparent remains a very appropriate role for a project manager.

Excellent project managers do not remain in the role of project manager for a long period of time. Excellent project managers are seen as individuals who understand not only the process but also how to get things done. These individuals are encouraged to continue using their skills in higher and higher roles in the organization. In fact, many organizations use the project management process to see how a new leader will actually do.

These organizations see project management as a good way of showing talented individuals their strengths and weaknesses, as well as helping them understand what lies around the next turn on their career path. A project manager must deal with personnel and budgets as well as the complete political realm of the organization. Therefore, a project manager not only is well placed to learn how the organization works, but also becomes a candidate to run the organization in the future.

The Black Knight

On occasion, project managers need to come in and do the dirty work. A team might not be working out, and may need to be restructured. Other times, a project will be deemed unviable by the higher-ups…and the project manager must come in to bring the project to a close.

This makes the project manager like the Black Knight—the stereotypical "bad guy" in medieval dramas. Project management is a relatively easy task when everything goes right…but the role becomes much more difficult in the face of an unpopular change. This is where you must have well-developed "people skills" in order to do what needs to be done tactfully and honestly.

Occasionally, a project manager must remove a team member from the round table. This is one of the least popular parts of project management experience, but it does happen and is part of life. The best advice is to get it done above board and quickly—like ripping off a Band-Aid.

Leadership Exercises: 50 Percent More Brain, 50 Percent Less Storm

Most employees will react to team exercises and mission statements as if they are "just the flavor of the week"—which is natural, since in many businesses these exercises do become the flavor of the week. The cycle feeds itself: employees are skeptical about a new exercise and don't put their all into the activity or the results. Management gets frustrated that the exercise hasn't run as smoothly as they expected, or hasn't paid off the way they expected, and so they scrap the system. When management is inspired by a new seminar or outing and wants to try another "sure-fire" activity, employees are even more skeptical, given the first failure, and so try even less. And the whole process feeds into itself.…

In the "Team Exercises" section, I shared some practical things one can do to encourage participation and integrate activities into the team's existing schedule. In addition to those suggestions, you might consider adopting a new series of activities for yourself. Consider the activities here to be "team exercises for groups of one."

By changing the way you do things personally, you set an example to other team members and encourage them to try new methods. You might also find yourself becoming more organized and displaying better leadership qualities.

These exercises are designed to be integrated into daily routines that already exist—many of them are purely mental exercises. Take five minutes at the start of each day (or whenever you are least likely to be interrupted) to try one. Once you have made one of these activities a routine, work on the next.

The Twenty-to-Breakthrough List

Good ideas are everywhere; so are bad ideas. Unfortunately, the obsessions with brainstorming and "thinking outside the box" common in the 1980s and 1990s didn't leave much of a distinction between the two. Don't get me wrong; brainstorming isn't a bad thing. But if you want your brainstorming to eventually get you to a really good idea, you have to do a lot of it.

Business experts and psychologists estimate that it takes about twenty ideas generated to guarantee that one or two of those ideas will be worthwhile. So get into the practice of generating at least twenty ideas for any project, activity, or framework.

For example, if your team is trying to think of ways to do low-cost marketing, don't stop exploring ideas until you can list twenty of them. Have a great idea for a back-end product? Just to be sure, try thinking up twenty product ideas. What you will find is that the first eight to ten ideas will come rather easily...but the next ten will take a lot of creativity, a little research, and a whole lot of effort. But learning how to break through this "ten-idea barrier" is crucial to the whole exercise. Once you are used to generating twenty ideas at a time, you will find that good ideas come more and more easily—and you'll be much better at spotting bad ideas as well.

In One Word

Few would doubt that communication is critical to good project leadership; indeed, pointing this out in print seems almost trite or redundant. But how often do we really practice the art of clear, concise communication?

Can you sum up what your team does in one sentence? Can you sum up what your team does in one word?

The challenge that most people have with communicating doesn't have to do with being brief (although lack of brevity is a common problem). Rather, the problem has to do with a lack of specificity. When we summarize, our tendency is to talk in generalities, or to create artificial-sounding phrases or buzzwords. But rather than clarify our message, these tendencies detract from our main point.

The first challenge, then, is to expand your vocabulary. There are many ways to do this: read more magazines and books, particularly ones aimed at a more literate audience; get a word-a-day calendar; and do vocabulary-building exercises at your desk. Expand your vocabulary as much as you can, and practice new words whenever you get the chance.

Whenever you find new words that might apply to your team, your process, or the image you want to project, make note of them and their definition. Keep the list accessible so that you can refer to those words when you need them.

The second challenge is to use that vocabulary in your communications, and do so in a way that is brief and precise. Practice brevity by cutting down every communication you send out. If you are sending out a memo that is one page long, see if you can pare down that memo to half a page. (Once you have some practice, you will want to further shrink that half-page memo to just four sentences!) Does your organization or product have a catch-phrase or jingle? Find a way to cut that phrase or jingle in half. Go through the employee handbook and look for ways to summarize all the important points in just a three-page "cheat sheet."

As you work on these challenges, try to keep the following guidelines that professional scholars and writers follow in mind:

1. *Avoid clichés and colloquial turns of phrase.* They add little to the meaning of what you say or write.

2. *Use more verbs and fewer nouns.* Saying "We implemented the process" is much clearer than "Implementation of the process was done" or similar expressions.

3. *Following no. 2, resist the temptation to "nounify" words.* You can do this by avoiding, as far as you can, words that end in "-ation," "-ition," "-ize," and "-ity."

4. *Try to avoid using the verb "to be" and all its conjugations (is, was, were, will be, etc.) in everything you write.*

5. *When you do use nouns, try to use concrete nouns as much as you can.* A simple rule of thumb: when you use a noun, ask yourself, "Is that something I could pick up with a fork or shovel?" If the answer is yes, fine. If the answer is no, you might be dealing with a noun that names something abstract, or just something hard to imagine. Try to find the concrete manifestation of the abstract thing you are talking about.

6. *Cut transition words—like "usually," "always," "however," "in case," and "of course"—down to a minimum.*

Admittedly, these rules are very difficult to put into practice—in fact, this book routinely violates them! But for an informal book-length treatment, violations of the rules are fine. For brief memos, e-mails, and other vital process communications, following the rules strictly will improve communication dramatically.

Recall the quip that an editor made to an author: "I like everything about your book but the ending," he said.

"What's wrong with the ending?" asked the author.

The editor's response: "It needs to be closer to the beginning."

When you have finally gotten into the practice of good, clear communication, do exactly what this activity suggests: try to sum up what, exactly, your team does in just one word. How's that for clear and precise?

Looking in Unlikely Places

Information is all around us, and most organizations work at putting information in the right place—or in the hands of the right people. But oftentimes innovation comes from looking for information or ideas in unlikely places. Cutting disciplinary boundaries should be done with caution, but the careful observer can reap the benefits of "dipping" into other areas.

Robert Wider, a creativity expert, put this idea across most clearly. He said, "Anyone can look for fashion in a boutique or history in a museum. The creative explorer looks for history in a hardware store and fashion in an airport." If you dedicate your mind to a project, you can begin seeing the world in terms of your project and the challenges your team faces. By keeping your eyes open, you might find solutions in unlikely places.

For example, World War I military designers looked to the Cubist paintings of Pablo Picasso for ideas about efficient camouflage patterns. The Wright Brothers learned most of the principles of aerodynamics by racing bicycles. Mahatma Gandhi's ideas about passive resistance came from the beatitudes in the Gospel of Matthew. Brain-imaging technology was originally used to calculate the density and brightness of parts of the sun. Velcro was inspired by the problems that gardeners were having with the common burr. Try to make a daily habit of looking into others' projects, areas, and fields for information and inspiration.

Trying the Unexpected

In general, project leadership requires vision, organization, and commitment—or, in other words, stability. Stability is good, and should be practiced and maintained when possible. But when faced with a challenge, sometimes the best strategy is to come from a completely unexpected angle.

Case in point: In 1334, the Duchess of Tyrol besieged a small castle belonging to a local lord. After months of siege, the defenders were out of supplies. Their last bit of food was a single ox.

The duchess, too, was having problems. She also was running out of supplies and had trouble maintaining discipline amongst her ranks. Tiring of the situation, she sought a way to end the siege quickly.

The defenders seized upon this situation: they loaded their last ox on a catapult and heaved the bovine over the castle wall—a move that surely seemed ludicrous at the time. The duchess, however, took this as a message: "We have enough food that we can waste our cattle. Do you?" This one act immediately discouraged the duchess and made her give up the siege. Because the castle defenders had done the opposite of what she expected, they managed her impression of the situation to their advantage.

Managing impressions sometimes requires an understanding of what is commonly expected of us, and then doing something else. Try the unexpected every once in a while in situations where people's common impressions might need retweaking.

Using Metaphors

We have all been warned about the dangers of "comparing apples and oranges." Granted, these are two different fruits. But, if you think about the comparison for a while, you'll realize that these seemingly dissimilar things actually have a number of attributes in common: they are both fruits; they are both round and roughly baseball sized; they both have seeds, skin, and a stem; and so on.

Take any two objects in the universe, and there will be infinite ways in which they are different—and infinite ways in which they are similar. Exercise your ability to see similarities by fleshing out metaphors when you can.

For example, how is a cell phone like a dog? Both make loud noises to get our attention. Both come in a number of varieties and kinds. Both can go with us on walks. Both can be given "commands" of sorts. Both can be obtained fairly cheaply, but the monthly maintenance cost can be a killer.

You get the idea. Now try applying the power of metaphor to the processes and principles of your organization. How is what you do like running for elected office? Or conducting an orchestra? Or assembling a watch? Or painting a work of art? Look for similarities and comparisons between the things that your team does, and other things that don't seem similar at first.

Read Your History

Project leaders are entrepreneurs in their own right. But every great entrepreneur does his or her research before diving in with a business plan. To be effective, a good project leader should know the history of his or her project and of similar projects.

What if there is no history of your project? What if you are trying something new? Or simply want to get away from the tired old ruts of what others have tried? Then read the history of other organizations. Though the details might differ, the principles behind good leadership (and bad leadership) are remarkably consistent. Read history to find the patterns.

Mao Tse-Tung's guerilla warfare in China provides a model for viral marketing. The British Empire is an excellent example of distribution and outsourcing. Paul Revere is proof of the need for networking before an important message needs to get out. The Roman Empire is a model of a business hierarchy—and the politics, deceit, and outright civil war that come with the hierarchy.

Take a few minutes a day to mine various histories. What can you use from your own personal history? The history of your teammates? The history of your project or organization? The history of your country, your culture, or the world?

Letting Chance Have a Role

Anyone who has tried to manage has probably stuck by the saying "Leave nothing to chance." Nonetheless, chance can be a good influence on us when our decisions and practices are mired in dogma and habit.

There are many ways to "consult" chance: tea leaves, dice, opening a book at random, reading horoscopes, and the like. (Note: I'm not going to argue that there is something mystical going on with these methods!) Find a method that seems comfortable to you, and use that method to stimulate thinking in different ways.

There's an old story about an entrepreneur who grew a single fishing boat into a whole fleet of boats and his own fishing company. When times were lean, his boat captains would come to him for the "latest maps" on new fish catches. He would go into his office, take a map of the coast, and crumple the paper. He would then uncrumple the paper and mark in some of the creases in the water parts of the map with a small blue marker. After making a few photocopies, he'd give the maps to his captains with the new "fishing lanes"—and the fleet would be wildly successful.

The lines were random, of course. The entrepreneur did not know whether his lines indicated new populations of fish or not. The point was that he got his captains to fish in new areas instead of overfishing areas that were already used up. He essentially pointed his captains to places that they had looked over.

Try introducing a random method or two into your routines. Use an external source for the randomness so you won't be biased or tempted to cheat. See where randomness takes you.

Ask Questions Differently

Questions are not innocent: the way we ask questions often influences the kinds of answers we get.

My favorite example comes from the Space Race. In space there is no gravity, which creates a number of challenges for astronauts. One simple but annoying challenge was writing in space. A standard ball-point or fountain pen needs gravity to move the ink to the end of the pen. No gravity, and the pen does not work.

NASA ended up spending millions of today's dollars on the problem. They hired a team of engineers to look at pens and try design after design until they had one that worked in zero gravity. The project worked, and the result was the space pen (which is still available as a gift shop novelty today). In essence, NASA used brute force (that is, labor and money) to answer the question "How can we get a pen to work in outer space?"

The Russian space program, on the other hand, did not have the money or the manpower to throw at the problem, and yet they found a much simpler and more elegant solution.

They decided to use pencils.

While NASA asked, "How can we get a pen to work in space?" the Russian space program asked, "How can we allow our astronauts to write in space?" Because the question asked was different, the solution was different as well.

Try asking questions in different ways and see if the answers differ. If you get stuck for a solution to a problem, take a stab at reformulating the problem. Try having your team approach challenges from different viewpoints by finding different ways of asking the same question.

Let Go and Let Nature Take Its Course

The biggest risk that a project manager faces is micromanagement. This is natural: if you are communicating the vision, making the plans, and organizing the team, you probably have a very strong sense of what needs to get done, and how things should look by the end.

But remember that you are not all-knowing. Sometimes the right thing to do is to sit back and let problems solve themselves in a more organic matter.

Take the story of the two architects charged with designing a college campus. The older architect decided to let the younger architect plan the landscaping and walkways as an exercise. The younger architect tried plan after plan, but his older colleague found problems with all of them. Finally the veteran said, "Look, just have the landscape crew plant solid grass across the campus, then let's reconvene a month after the semester starts."

The crew planted the grass, from door to door, with no walkways or patios. As the semester started, students simply cut across the quad, taking the most natural and efficient paths for them. They also congregated at convenient gathering points in the quad. One month into the semester, the quad was crisscrossed with areas of downtrodden grass. The senior architect then told the landscape crew to pave the paths that the students naturally made.

The result: The paved paths were placed exactly where students needed them, because they were based on the natural traffic patterns of the college, and the patios were placed in areas where students naturally tended to congregate. Not only was the layout optimally useful, but also the designs thus made had a strange beauty of their own.

What are you trying to force? What are you designing, managing, or imagining according to your own whims instead of going with a solution that is natural for the team? Try "letting go" of some of the processes and details you are focusing on, and see what happens.

Developing Decision-Making Skills

How can decision-making skills be developed? Using the U.S. Marines' philosophy of simplicity, here are some steps to improve the decision-making process. Following a simple step-by-step approach to decision making can increase your ability to make good decisions.

Making good decisions will do the following:

- Build the confidence needed to make more decisions
- Help you achieve goals
- Help you avoid mistakes

Step 1: Define the Decision

Without being concerned about the cause or solution temporarily, define the problem to be resolved or decided.

- Put in words exactly what the problem is.
- Write a description of the problem as if you were going to hand it off to someone else for a solution.
- Collect any knowledge or specific information that could affect your decision.
- Get input from the people who will be affected by your decision.
- Get input from someone who will not be affected by your decision.

Step 2: Imagine There Is No Decision to Be Made

- Use your imagination to surmise how things would be without the problem.

Step 3: Imagine Doing Nothing

- Imagine the outcome if no decision is made.
- Write down what would happen if you do not make any decision.

Step 4: Create the Perfect Outcome

- Imagine the perfect outcome: if you had the ability to wave a wand and solve the issue, what would the outcome be?

Step 5: Imagine Decisions

- Is this a 50-50 decision? (Yes or no.) If so, imagine the outcome of both decisions.
- Is there a multitude of solutions to choose from?

- Imagine the perfect decision to the problem—even if it is unrealistic. For example, if the problem requires money—more than you have available—imagine you have an unlimited bank account for this decision.
- Consider alternative decisions.

Step 6: Narrow Down Solutions

- State the goal of your decision.
- Now make a list of as many possible realistic solutions you can think of that will achieve the objective of the decision.

Step 7: Consider Your Resources

- Consider the resources your decision will require—time, energy, equipment, and money.
- Pick the best three solutions (the rule of three). Consider logic and intuition.

Step 8: Consider the Timing

- How long do you have to make your decision?
- Will the outcome be different based on when you make your decision?

Step 9: Gather Input

- Get input again from the people who will be affected by your decision.
- Get input again from someone who will not be affected by your decision.

Step 10: Take Action

- Inform the people involved of your decision.
- Record the outcome of your decision and file with the work for future reference.

For a template and other materials, please consult our companion website, www.ManagementToLeadership.com.

Glossary

Accountability: Setting standards, upholding those standards, and helping people become the best they can be day in and day out.

Authority: Created through influence, by using techniques that make people want to work with and for you; grows out of trust, competence, and a reputation as a problem solver.

CEOS: Refers to an organization's relationship with their Customers, Employees, Owners, and Significant others.

Conflict: An essential element of a high-performing team; conflict is used as an exploratory process, and a way for team members to challenge one another to defend their beliefs and assumptions in order to make a decision based on the best interests of the team and project.

Discipline: A corrective action that requires the leader to identify actions that are not in line with the agreed-upon expectation and demonstrate to the offender what the proper action should have been.

Empowerment: Enabling employees to make decisions for themselves, resolve problems, and take responsibility for their actions based on the organization's mission, values, and goals.

Fundamental attribution error: The tendency of individuals to falsely attribute the negative behaviors of others to their character while attributing their own negative behaviors to their environment.

Generation Y: An emerging generation of highly talented and efficient workers born between the years of 1980 and 1995, Generation Y is 60 million strong and makes up approximately 22 percent of the workforce.

IDEAS: An acronym used to describe the requirements of true teamwork: Identifying attributes, Debating essential issues, Embracing accountability, Achieving commitment, and Setting and maintaining high standards.

Influence: The ability to get others to do your will because they want to follow you, not because of your position or power over them.

Organizational clarity: Defining and agreeing on the fundamental ideas behind an organization and effectively communicating those ideas to everyone within the company; gives employees at all levels a common vocabulary and a common aim.

Power: The ability to force people to do your will because of your position; using threats, punishment, and bribery to coerce people to complete tasks.

Project charter: A starting document that defines the project, identifies the project's status, and explains how the project will be completed; derived from the original business plan.

Servant leadership: Developed by Robert K. Greenleaf in 1970, a leadership model that proposes that a leader must serve the people who follow, viewing them as an end in themselves rather than a means to a corporate purpose.

Bibliography

Allen, M. 2004. "Leading from behind." *Dallas Business Journal*, November 26.

Baker, P. 2005. *Employer Secrets (and How to Use Them to Get the Job and Pay You Want)*. Saint John, NB: Dreamcatcher.

Datasegment.com. "Commitment." http://onlinedictionary.datasegment.com/word/commitment

Greenleaf, R. 2002. *Servant Leadership: A Journey into the Nature of Legitimate Power and Leadership*. New York: Paulist.

Hill, L. 2008. "Where will we find tomorrow's leaders?" *Harvard Business Review*. http://harvardbusiness.org/product/where-will-we-find-tomorrow-s-leaders-a-conversati/an/R0801J-PDF-ENG.

Kohn, A. 1999. *Punished by Rewards: The Trouble with the Gold Star, Incentive Plans, A's, and Other Bribes*. New York: Mariner, 1999.

Lencioni, P. 2005. *Overcoming Five Dysfunctions of the Team*. San Francisco: Jossey-Bass.

———. 2000. *The Four Obsessions of an Extraordinary Executive*. San Francisco: Jossey-Bass.

Mandela, N. 1995. *Long Walk to Freedom*. Boston: Little, Brown.

Maxwell, J. 2006. *The 360 Leader: Developing Your Influence from Anywhere in the Organization*. Nashville, Tenn.: Nelson Impact.

Spears L. C., and M. Lawrence. 2004. *Practicing Servant Leadership: Succeeding through Trust, Bravery, and Forgiveness*. San Francisco: Jossey-Bass.

Index

For Product Safety Concerns and Information please contact our EU
representative GPSR@taylorandfrancis.com
Taylor & Francis Verlag GmbH, Kaufingerstraße 24, 80331 München, Germany